Penguin Education
Modern Science Studies

Modern Biology

Edited by Roland Hoste

Modern Biology

Selected Readings

Edited by Roland Hoste

Penguin Books

Penguin Books Ltd, Harmondsworth,
Middlesex, England
Penguin Books Inc., 7110 Ambassador Road,
Baltimore, Md 21207, USA
Penguin Books Australia Ltd,
Ringwood, Victoria, Australia

First published 1971
This selection copyright © Roland Hoste, 1971
Introduction and notes copyright © Roland Hoste, 1971

Made and printed in Great Britain by
Richard Clay (The Chaucer Press) Ltd,
Bungay, Suffolk
Set in Monotype Times

Contents

Part Three
Structures, Processes and Control in Populations 229

Introduction

In this book of Readings I have attempted to build a bridge between the basic biology in school syllabuses and the activities of research biologists working in the pure and applied fields. The book has been divided into three sections dealing with cells, organisms and populations because each of these levels of organization has an organized, integrated unity of its own.

Each section should show the way in which the structure, whether of cell, organism or population, can maintain itself in a fairly steady state as a result of adjustments within itself, a process known as homeostasis. It should be possible to see how the parts of the structure have evolved in form and function so that they contribute to homeostasis.

At the cellular level a main preoccupation is with the transformation of energy into a form which can be used by living organisms, and the control of these 'biochemical factories' by genes. The behaviour of well-regulated cells contributes greatly to the health of the organism. These, in turn, must regulate their activities so that the appropriate behaviour, or oxygen concentration occurs in the right place at the right time.

Part Two, *Processes and their Control in the Organism*, illustrates a variety of homeostatic mechanisms at work within organisms. We are now concerned with the interrelationships of the various organ systems within the animal or plant, and how these interrelationships are maintained by chemical means, or by a nervous system. On occasion, there is no need for communication between the systems: the information essential for the functioning of the system has been incorporated within them as a result of the evolutionary process, and the structure is so adapted that it is an integrated unit. In other instances the information has not been there from the beginning, and the organism has to adapt itself, that is, it must learn to change its behaviour.

Part Three, *Structures, Processes and Control in Populations*, illustrates homeostasis in populations by means of a number of Readings concerned with animal populations and then shows how

the human population regulates itself. Here, the interest is in how a population regulates its numbers, and how it adapts itself by altering its genetic structure. It is also possible to compare the ways natural populations regulate themselves with the methods of human populations.

Each section of the book consists of a collection of papers concerned with some aspect of the biology of cells, organisms or populations. Some are articles from 'popular' journals which present a survey of a topic presented in relatively straightforward language. Others are review papers designed for specialists, and some are reports of original research which are intended to communicate the findings to other research workers.

This selection of Readings represents a considerable range of difficulty from the popular article to the complex review paper. The latter have been prefaced with a short note explaining the background to some of the specific topics raised. These papers often require a detailed knowledge of the topic and are sometimes couched in technical language leading to a rather tedious style.

To make the most use of these papers, they should be digested slowly. The unfamiliar vocabulary should be learned, and the evidence and arguments examined critically. Most original research papers begin with a summary of their contents, and often have an introduction giving an outline of the present knowledge of the topic, and quoting the evidence found by other workers supporting or contradicting the paper's theory.

Part One Cellular Production Lines

Cells in a multicellular organism are much more than the blocks of which it is built. They are complex structures which play a vital role in the life of the organism, and (as Lehninger shows in Reading 2) can be considered as analogous to factories within the organism. In these factories are production lines which have specific tasks including:

1. liberating energy which may be used either on another production line within the cell or elsewhere in the organism;

2. constructing molecules which may be intermediate compounds forming the raw materials for another production line within the cell. Alternatively, they may be final products for use either within the cell (e.g. photosynthetic pigments) or outside the cell (e.g. hormones);

3. helping the cell to provide a service, e.g. contraction in the case of muscle cells.

All production lines require basic raw materials from outside the factory, and the entry of these must be regulated, as must the dispatch of manufactured products and the disposal of waste materials. In the cell all these functions are carried out by the outer membrane, and its structure and physiology must be such that it is able to do this. The production lines within the cell are located on the internal membrane system, the endoplasmic reticulum, at sites organized to allow particular processes to go on at optimum rates.

Research into the structure of the membrane is helped by a knowledge of the processes which go on at a particular site, and understanding of membrane structure permits a greater

appreciation of the processes involved. We are, of course, nowhere near a full understanding of membrane structure, nor of cellular biochemistry, but gradually the jigsaw of evidence is being fitted together. In Reading 1, Chapman discusses some of the evidence which has accumulated in various scientific disciplines concerning the structure of cellular membranes. He demonstrates how important a good knowledge of physics and chemistry is to the solution of biological problems, and gives an indication of the strategy which will be used to fill some of the outstanding gaps in our understanding.

Two other important points arise in Chapman's paper. One is the dependence of advances in scientific understanding upon advances in technology. Little could be known about cellular membranes until advances in electron-microscopy techniques made the observations possible. Secondly, the same evidence can be interpreted in different ways by different scientists, depending on their own backgrounds.

In Reading 2 Lehninger (who developed the idea of cells as factories) develops the important ideas behind the energy relationships of a cell. Energy has to be converted into a usable form. In many factory engines this form is heat, but living cells use chemical energy in the form of an 'energy-rich' compound, adenosine triphosphate (ATP), which contains a high-energy phosphate bond. In the part of Lehninger's article reprinted here (the rest of it contains a fuller account of the energy relationships of a cell) he describes the high-energy phosphate-bond concept, and shows how ATP is used to drive two of the energy-requiring systems in cells, the synthesis of large molecules, and the transport of substances against a concentration gradient.

The ultimate source of the energy used in cellular processes is the sun. Energy arrives on Earth as a wide spectrum of radiations from the sun, and part of this spectrum is trapped in plants in the process of photosynthesis. It is then stored in complex molecules of starch, sugars, etc., to be used later by the plant in the reactions described in the previous article. If the plant is eaten by a herbivorous animal, the stored energy may be used by the animal instead. But the herbivore may also

be eaten by a carnivore before that energy has been utilized. Thus it can be seen that photosynthesis is essential for all life, not just that of plants, and consequently it is one of the most important processes for a biologist to study.

But how did this process start? Obviously we cannot know, but the next two papers throw some light on this question. Echlin, in Reading 3 *Origins of Photosynthesis*, follows clues provided by studies of present-day blue-green algae and photosynthetic bacteria, which have a long geological history. He asks the question 'Are the chloroplasts of modern higher plants the descendants of primitive blue-green algae?' Is it possible that any other structures within cells of higher organisms could perhaps be derived from parasitic, or symbiotic unicellular creatures? In Reading 4 Chedd considers evidence from a different line of investigation. Life may have originated on this planet, or it may have started elsewhere and been carried here by extraterrestrial objects like meteorites. Several experiments seem to show that, given the atmospheric conditions which are thought to have existed four billion years ago, something with many of the attributes of life could have arisen. Chedd reports experiments which attempt to recreate this primitive atmosphere – and in it molecules arise which could have been precursors of living structures. One of these experiments results in the production of porphyrins, substances able to absorb energy in the visible spectrum: the significance of this will be apparent after reading Echlin's article.

Reading 5 by Rogers and Goodwin gives an account of experiments which have been carried out to study how a green plant forms its own photosynthetic pigments. The membrane of the living cell is important in the regulation of synthetic activity by controlling the passage of substances from one part of the cell to another. The permeability of the membranes is thought to be governed by phytochrome.

At a more fundamental level, control of cellular activity is accomplished by the genes which are mainly to be found on the chromosomes, but many are known outside the nucleus. The way in which genes exert their control is comparatively

well known now, thanks to the impetus given to this branch of biology by the discovery of the structure of DNA (deoxyribonucleic acid). This brilliant piece of deduction by Watson and Crick rapidly led to the cracking of the genetic code by which the blueprint for protein synthesis is carried in the cell. The proteins manufactured are often enzymes which catalyse cellular activities outlined at the beginning of this introduction.

But what controls the genes? Every cell contains a full complement of genes, but only a few are operative at any one time: the genes in action during pupal stages of an insect may not be required in the larva nor in the adult. Some may not be used at all; for example, the genes which are responsible for the development of female characteristics which are present, but inoperative, in each male cell. What switches them on and off? Readings 6 and 7 are concerned with the fundamental system (or systems) which control the genes. As in most topics at the 'frontiers of knowledge' there is very little agreement among the experts. Both authors discuss some of the rival hypotheses which have been put forward to attempt to explain the phenomenon, and, as Chedd hints in Reading 7, we may have reached a point where yet another hypothesis is required before we can resolve all the conflicting evidence. Both Chedd and Butler point out the link between this theoretical problem and the practical one of alleviating the group of diseases known as cancer which seems to result from a cell losing control of its genes. Butler's paper also provides a brief outline of the system by which genes are thought to control the synthesis of protein.

Cells not only build up proteins and other large molecules, but also break them down. It seems that the enzyme systems responsible, the lysosomes, are located in the cytoplasmic fraction of a centrifuged cell-suspension (the heaviest portion of the mixture is forced to the bottom of the centrifuge tube whilst the lighter portions remain above it, so different parts of the cell can be separated physically from each other). Each part can be identified on the basis of its behaviour, or its physical form, or its chemical composition. Reading 8 by Afzelius shows how these characteristics were used to track

down the lysosome. But it was not certain that the chemically defined lysosome was the same as the physical structure revealed by the electron microscope.

These are not the only problems with lysosomes, which illustrate many of the difficulties of cellular biology, where the evidence is often, at best, circumstantial. Afzelius raises several of them: many have not yet been resolved.

Reading 9 by North shows how the evidence for the structure of protein molecules was gathered. The enzyme he deals with is lysozyme (not to be confused with the lysosomes discussed in the previous reading), which is responsible for the lysis of certain bacterial cell walls, and is the first enzyme to be analysed completely. Evidence comes from several scientific disciplines and is pulled together by someone who understands the living organism and the physical and chemical nature of its environment.

The following list explains the major cell components and their probable functions.

Structure	Probable function
Nucleus	Synthesis of DNA and messenger RNA
Nuclear membrane	Controls entrance and exit of materials to and from nucleus
Endoplasmic reticulum	Involved in many cellular processes
Cell membrane	Controls passage of material in and out of the cell
Mitochondria	Oxidative phosphorylation and amino and fatty acid degradation and synthesis
Ribosomes	Protein synthesis
Lysosomes	Serve to digest proteins, carbohydrates and nucleic acids
Golgi apparatus	Involved in synthesis of fats
Centrioles	Involved in cell division

The papers in this part cannot cover all aspects of cellular biology and a good overall picture may be obtained by reading parts of the books listed in Further reading on page 345.

1 D. Chapman

Biological Membranes

D. Chapman, 'Biological membranes', *Science Journal*, vol. 4, 1968, no. 3, pp. 55–60.

The search for information on the structure of biological membranes has received considerable impetus from recent advances in molecular biology. As more is learned about the detailed structure and organization of individual cellular components, such as protein and nucleic acids, attention naturally turns to the organization of the cell itself – a province of biological organization dominated by the cell membranes. Membranes are implicated in the storage of information, the influence of drugs, the transport of materials, the scavenging of pathogens and the organization of a variety of energy-transfer systems to name but a few key roles. It is the vital biological interface in terms of cellular activity and it is the living organism's contact with its external environment.

For many years the idea of a cell membrane existed almost entirely at a conceptual level; physiological experiments suggested a barrier to the exchange of material between the cell and its surroundings. It was called a membrane because the barrier was thought to be a thin layer completely enclosing the cell contents. The membrane appeared to have definite mechanical and physical properties and, even in the absence of any direct analyses, suggestions were made as to its structure and organization. With the advent of the electron microscope, the problem appeared to be solved. Electron micrographs apparently confirmed the existence of the outer cell membrane and gave support to the theory that a bilayer of lipid (fat) molecules sandwiched between layers of protein molecule was the important structural unit. Inside the cell, structures called mitochondria and lysozymes were discovered which were seen to be bounded by thin, layered structures or membranes. Paired membranes called the endoplasmic reticulum were also observed. Similar structures were also observed

with the chloroplast of plant cells and in the rods and cones of the eye.

Recently, the detailed structure of cell membranes has suddenly become a live issue again as a result of new techniques and new ways of thinking. According to Aldous Huxley, 'Man makes God in his own image, there are many men, therefore there are many Gods.' The same can now be said of theories concerning the structure of biological membranes. There are almost as many theoretical models as there are scientists concerned with their study. At times attitudes apparently reflect the particular expertise of the scientists concerned. Thus lipid chemists often see the membrane as being dominated in structure and function by the lipid material while the protein chemists credit protein with the dominant role.

Unlike the pioneer students of membranes the current protagonists have a wide range of sophisticated instruments and techniques at their disposal some of which are only now being applied to biological material. For example, our work at the Unilever Research Laboratory, Welwyn, involves the construction of model membranes as well as the analysis of biological membranes using spectroscopic techniques. We are no longer restricted to the indirect evidence for the membrane; we can examine directly its molecular structure and in this way hope to learn more about the association and interaction of its chemical components. With such prospects, it is hardly surprising that the study of biological membranes has become an exciting and important area of contemporary research.

To understand the structure of biological membranes at the molecular level, their various components must be identified and isolated in a pure state. This is by no means easy, particularly because the membrane preparation must lack none of the membrane components themselves. The problem of isolating pure biological membrane material therefore involves: establishing a method for destroying cells; isolating the membrane material; demonstrating the material's precise origin; and, finally, estimating the purity of the preparations.

The biological membrane which has received considerable attention is that of the erythrocyte – the red blood cell. When

erythrocytes are subjected to certain osmotic pressures or enzymes, they undergo haemolysis – a process in which the cell membrane structure loosens sufficiently to allow the cell contents to escape without itself being ruptured completely. All that remains are the 'ghosts' of the original cells – a collection of erythrocyte membranes.

The success of the isolation of the membrane material depends also on its identification. Electron microscopy is sometimes useful. Surface antigens which attach to specific target chemicals have been used as marker materials. A difficulty in this approach is to ensure that the marker molecule is specific to the membrane. Attempts have been made to overcome this by artificially labelling the membrane with a chemical known to be confined to the membrane and easily detectable during the separation of the membranes from the constituents of the disrupted cells.

To appreciate which materials are contaminants of this membrane fraction is also rather difficult. With erythrocyte ghosts, the question has been raised as to whether haemoglobin is a true component of the membrane or whether it is a contaminant. This question is still unresolved and the significance of the part played by haemoglobin in the membrane structure is somewhat obscure.

Recently the modern techniques of lipid analysis have been increasingly applied to membrane preparations. The membrane materials which have been analysed appear to contain both lipid, protein and carbohydrate material. For example the red blood cell membrane contains 35–45 per cent of lipid, 50–60 per cent of protein and some 8–10 per cent of carbohydrate. Other important components are water and metal ions.

As early as the 1930s, the biological membrane was pictured as a bilayer of fatty molecules sandwiched between protein layers. The fatty phospholipid molecules were found to consist of two parts – hydrocarbon chains of carbon and hydrogen atoms attached to a complex structure made up of carbon, hydrogen, nitrogen, oxygen and phosphorus atoms. The two components in turn endow the phospholipid molecules with a contrasting polarity. The carbon–hydrogen chain is soluble in organic solvents such as ether but is insoluble in water, whereas the other part of the molecule is soluble only in water. In other words the

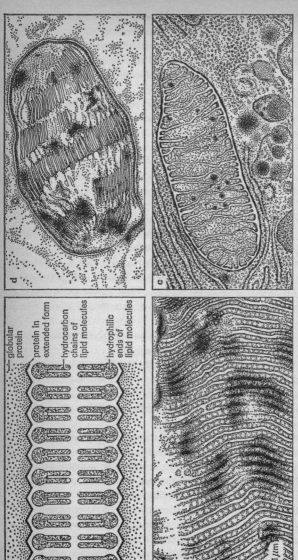

Figure 1 The classical structure of the membrane shows two layers of lipid molecules sandwiched between protein layers (a). The hydrophilic heads of the lipids lie next to the protein. It has been suggested that there is a 'unit structure' for all membranes such as those in the illustrations of the retinal rod of a perch eye (b), mitochondrion from a bat pancreas acinar cell (c) and a tobacco-leaf chloroplast (d). Recent studies raise the question as to whether the classical structure of the membrane is correct

Labels in figure (a):
- globular protein
- protein in extended form
- hydrocarbon chains of lipid molecules
- hydrophilic ends of lipid molecules

Scale bar (b): 0·1 μm

phospholipid molecule has a hydrophobic (water-repellent) group and a hydrophilic (attracted to water) group. In the bilayer model the hydrophobic group is situated in the centre and the hydrophilic group lies on the outside next to the protein. Since this arrangement corresponds to a structure only a few molecules thick, it was clearly beyond the range of the light microscope. The search for such a structure by direct observation had to await the superior resolving power of the electron microscope.

With the advent of the electron microscope events moved rapidly. Electron micrographs of a whole host of cell membranes came under the scrutiny of the electron beam and artificial membranes composed of layers of phospholipid gave similar structures. The resulting membrane electron micrographs were interpreted to give further support to the suggestion that the bilayer of lipid is the important structural unit.

The similar appearance of the surfaces of many different types of cell, as well as the membranes of bodies such as the mitrochondria within them, led Robertson to suggest they were made up from a 'unit membrane'. In his preparations this appeared as a three-layered structure with two dense lines about 20 Å wide separated by a lighter area of 35 Å (one ångström unit is 10^{-8} centimetre). In his model of the unit membrane, Robertson went on to equate the dense lines with the proteins and hydrophilic polar groups and the lighter areas with the carbon–hydrogen chains. This concept appeared to combine the visual evidence of the electron micrographs with other chemical and physical evidence. By referring to the biological membrane as a unit, he emphasized that the three parts of the 75 Å structure were part of one membrane and also that all membranes had a similar molecular arrangement and origin. He was even able to suggest how the present structure of the cell might have evolved.

When the interpretation in molecular terms of electron-microscope data is pursued, however, a number of difficulties arise, not least of which are those stemming from the treatment of tissues preparatory to using the microscope. Before biological material can be observed with the electron microscope, it has to be 'fixed' using potassium permanganate ($KMnO_4$) or osmium tetroxide (OsO_4) to preserve the structure; it then has to be dehydrated, embedded in an appropriate medium and cut

into extremely thin sections. For the purpose of the present discussion the fixatives also act as 'stains' by combining with certain regions of the specimen. Their 'heavy atoms' are essential to interfere with the streams of electrons which bombard the specimen and thereby produce the image seen by the microscopist; most of the atoms which make up the membrane – carbon, hydrogen, oxygen and nitrogen – are comparatively light and contribute little to the final electron micrograph.

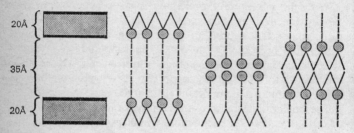

〜〜 protein

◎ hydrophilic ends of lipid molecules

–––––– hydrocarbon chains of lipid molecules

Figure 2 The unit pattern of membranes under the electron microscope appears as at far left. Next to this is shown the classical molecular structure for the membrane. The two right-hand diagrams show alternative structures which could also explain the observed unit pattern. Difficulties in interpreting electron micrographs account for the fact that no one structure is as yet universally accepted

Recently, it has been suggested that there is no way to interpret in molecular terms the dense lines seen in electron micrographs of membranes fixed with potassium permanganate, even though the triple-layered structure is usually distinctly demonstrated. The triple-layered structures are also often observed after fixation with osmium tetroxide but the location of the heavy atoms of osmium is disputed. Wigglesworth has suggested that the osmium attaches to the carbon–hydrogen chain of the phospholipid, while other workers claim that after a preliminary reaction at this site, the osmium derivatives migrate to the polar groups. Korn has voiced even stronger doubts concluding that at present the dense lines in membranes fixed with osmium tetroxide reveal

nothing about the molecular orientation of the lipids in the original membrane. He also points out that the triple-layered structure is also seen in fixed mitochondrial membranes from which *all* the lipid has been previously removed.

The location of the osmium and the nature of the action of potassium permanganate are not the only uncertainties in the interpretation of the electron-micrograph data. Practically every stage in the treatment of specimens, including the bombardment with electrons in the microscope, might be suspected of creating misleading artifacts.

A new technique introduced by Mühlethaler and co-workers has recently been introduced to electron microscopy which attempts to overcome the difficulties associated with the chemical fixation method. Known as freeze-etching, it depends upon freezing a small sample very rapidly, usually in the presence of glycerol, and then putting the specimen on to the cold stage of a vacuum coating unit. When a high vacuum has been obtained, the surface of the specimen is cut with a cold knife. The temperature of the specimen is then raised to $-100\,^{\circ}$C and a thin layer of ice is sublimed off. The cut surface is therefore etched by vacuum sublimation. Immediately after this the surface of the specimen is shadowed with a heavy metal and a carbon replica is made which is then examined in the electron microscope. The method involves no chemical treatment until the replica is formed.

A number of membrane types have now been examined in this way. The results broadly support the idea of a unit membrane and the technique appears to have important potential in the study of cell membranes. There are already, however, disagreements in some interpretations of the pictures obtained by this technique. Some workers suggest that freeze-etching may actually split membranes and reveal inner faces.

Two new spectroscopic techniques have also been recently applied to the study of membranes. The first is optical rotatory dispersion (ORD). The basis of this technique is as follows. Substances whose constituent atoms are not symmetrically arranged are optically active – they will rotate the plane of polarized light. The variation of optical rotation with wavelength is called optical rotatory dispersion. They also absorb left and

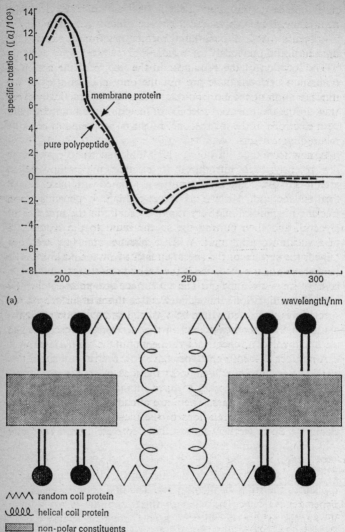

(a)

specific rotation $([\alpha]/10^3)$

wavelength/nm

membrane protein

pure polypeptide

ᐱᐱᐱᐱ random coil protein

ꝇꝇꝇꝇ helical coil protein

▨ non-polar constituents

● hydrophilic ends of lipid molecules

▬ hydrocarbon chains of lipid molecules

(b)

right circularly polarized light to different extents and thus exhibit the phenomenon called circular dichroism (CD).

CD and ORD spectra have become extremely valuable for elucidating the structure of soluble polypeptides and proteins. These techniques have now been applied by Wallach and other workers to the study of cell membranes in order to provide more information about their protein architecture. The picture that emerges is not easily reconcilable with the unit membrane type structure. For one thing, these studies indicate that the membrane proteins may be coiled up in an α-helix structure. It has been argued that the sequences of amino acids needed to form an α-helix protein are different from those required for a protein of the conformation required to fit into a unit membrane structure. One structure put forward to explain the ORD and CD spectra has an arrangement of lipid and protein such that the heads of the lipid molecules, together with all the ionic side chains of the protein, lie on the outside of the membrane. Those sequences of the protein which consist predominantly of non-polar side chains are in the interior of the membrane, together with the hydrocarbon tails of the phospholipids and the relatively non-polar lipids, such as cholesterol. In particular the helical portions of the protein are interior where they are stabilized by hydrophobic interactions.

Other workers suggest that suitable amino acid sequences could produce largely α-helical aggregates containing aqueous channels lined with polar side chains and that it may not be necessary to place all the ionic side chains at the membrane surface. It is argued that the native conformation of membrane proteins depends upon their apolar interaction with membrane lipids just as the conformation of haemoglobin is critically influenced by the hydrophobically bound haem groups. It has also been

Figure 3 The optical rotation of various membranes and their constituents varies with wavelength to provide a characteristic pattern. The graph (a) illustrates the specific rotation as a function of wavelength and shows the difference in spectrum associated with the membrane protein and, for comparison, a pure polypeplide. Note that the 237 nm trough for the protein occurs at longer wavelengths than with the polypeptide. This and other studies have suggested that some of the membrane protein may be coiled into an α-helix and that a hydrophobic lipid–protein interaction is occurring. The diagram (b) shows one way of presenting this information as a molecular structure

suggested that the membrane proteins are characterized by amino acid sequences which specifically adapt them to interact with certain lipid components of the membrane and the aqueous environment. Thus the amino acid sequence might determine the type of phospholipids present.

Recently in our laboratory at Welwyn we have applied another new physical technique to the study of membranes – the technique of nuclear magnetic resonance spectroscopy. A feature of n.m.r. spectroscopy is that it provides information about a particular nucleus. Thus it is possible to look at the hydrogen, phosphorus or nitrogen nucleus and obtain information about its degree of molecular freedom. In our laboratory we have recently used n.m.r. spectroscopy to study solid phospholipids in the presence and absence of water. We have shown that an increase of temperature produces a liquid crystalline, or mesomorphic, state in which considerable molecular motion of the polymethylene chains of the lipid occurs (see 'Liquid crystals', *Science Journal*, October 1965). At the same time the possibility of diffusion of the phospholipids becomes greater.

We have also carried out experiments with various types of membranes, for example red blood cells and the envelopes of *Halobacterium halobium*, and studied their n.m.r. spectra. With the red blood cell membranes signals are observed from the choline $^+H(CH_3)_3$ group from the phospholipid of the membrane. Other signals probably associated with sugar or sialic acid groups also appear. However, a signal from the hydrocarbon chains $(CH_2)_n$ of the phospholipid is absent. Only after we add materials which are known to disrupt hydrophobic bonds do the hydrocarbon-chain signals begin to appear in the n.m.r. spectrum. When trifluoroacetic acid is added to the membrane, signals also appear from the movement of the individual amino acids of the proteins which are now uncoiled.

These results suggest that the membrane protein is not in a condition which allows freedom of molecular movement of the individual amino acids until some disruptive effect has been applied to the membrane. Further, they suggest that the lipid hydrocarbon chain is not as free as it is in the isolated lipid. This may be because the hydrocarbon chains are interacting with the hydrophobic amino acids of the protein.

Despite present uncertainty of the details of biological membrane structure and the doubts about the nature of the main structural unit, several model membranes have recently been created which consist of individual phospholipid bilayers or, alternatively, of a concentric arrangement of phospholipid bilayers. An important advance was made in 1962 when Mueller and co-workers reported that, after painting phospholipid across an orifice, most of the lipid drains away to leave a film so thin that it can be considered to be bimolecular. The dimensions of this bilayer have been estimated on the basis of optical and electrical studies to be 60–90 Å thick. When protein obtained from various biological sources is added, the electrical resistance of these bilayers drops sharply. What has made this work particularly provocative was

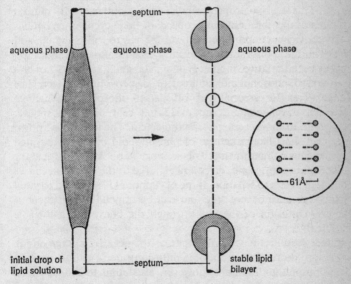

Figure 4 A membrane model can be prepared by painting phospholipid across an orifice. Most of the lipid then drains away to leave a thin layer of lipid believed to be bimolecular. The addition of biological protein to the thin layer gives the model an electrical activity which stimulates the behaviour of a nerve cell. Certain antibiotics increase the conductance of such layers by a factor of up to 10^3 and affect permeability to sodium and potassium ions

D. Chapman 27

the fact that this made the films electrically 'excitable': their resistance changes reversibly between two values in response to an applied voltage. It is thought that a precisely similar system causes a neurone in the brain to 'fire'. Indeed, the parameters of this model system has similarities to those of a frog nerve.

Recently, amongst the biological materials added to these black films have been some cyclic polypeptide antibiotics such as valinomycin, actin and enniatin. The absorption of these molecules into the black films increases the membrane conductance up to 10^3 times. The single ionic conductances are observed to differ by as much as 300 times and, most importantly, a *discrimination* between sodium and potassium ions is shown. A resting potential of up to 150 mV is given when 0·1 M solutions of sodium chloride and potassium chloride are placed on opposite sides of the bilayer. This is also a property similar to that obtained with natural biological membranes. This property of these cyclic polypeptides may provide an important clue as to the way in which sodium and potassium ion discrimination occurs in natural membranes.

In an alternative model system, Bangham and co-workers have used phospholipid globules or dispersions in water. The system consists essentially of a dispersed phospholipid, such as lecithin, which contains up to 15 per cent dicetylphosphoric acid and, in some cases, phosphatidic acid. The lecithin is known to form sheet-like structures of phospholipid bilayers separated by aqueous compartments. The presence of phosphatidic acid, or dicetylphosphoric acid, gives a net charge to the lipid and causes adjacent layers to separate. If the phospholipid is allowed to swell in the presence of ions, the ions can be trapped in the aqueous layers. Diffusion of the ions through the bilayers can then be studied.

Recent studies have shown that, as the negative charge on the structure decreases, so does the diffusion rate of cations leaving the phospholipid. Anions, however, are found to diffuse much more rapidly and appear to be relatively free to diffuse whether fixed charges of either sign are present or not. These globules have also been shown to have osmotic properties similar to those of cell membranes. They are also considered to exhibit other properties which simulate the properties of natural membrane systems;

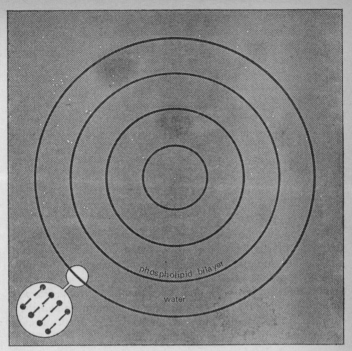

Figure 5 A globular model of membrane can be made by dispersing phospholipid in water. It then forms phospholipid bilayers, separated by aqueous compartments. Models of this kind are very useful for studying the diffusion of ions and other molecules through the layers

ion transport through these globules, for instance, is affected by steroids much as it is in natural membranes.

Recently in our laboratory we have been mimicking the electronic absorption spectrum of plant systems by incorporating chlorophyll into lipid globules of this type.

To summarize the present position we can say that the model put forward to describe biological membranes, based on the lipid bilayer structure, is still a strong candidate at the present time. With myelin membranes, X-ray and electron-microscope

evidence strongly supports this type of structure. However, experimental evidence coming from the studies of other membrane systems suggests that the bilayer model may not necessarily be the dominant structure. This new evidence suggests the basic feature of other membrane structures is the hydrophobic interactions between the lipid chains and some of the amino acids of the protein. Emphasis in these new ideas lays greater stress on the action of the protein, rather than the lipid, to explain the transport mechanisms of membranes. It also lays stress on the idea that the protein structure determines the lipid composition of the membrane rather than the reverse. More experiments need to be carried out to study and determine the association of the molecules of membranes. As yet little is known about the physical chemistry of the lipids and little about the way in which lipids and proteins can interact.

The model membrane systems are important and useful for testing hypotheses as to how various membrane processes occur, such as diffusion and the organization of energy-transfer molecules. The present mood of scientists working in the field is one of great optimism; there is the impression that considerable progress to our understanding of the structure of membranes and their function is being made, and that our knowledge of membranes will increase rapidly in the very near future.

2 A. L. Lehninger

The Transfer of Energy Within Cells: Cells as Factories

From A. L. Lehninger, 'Transfer of energy within cells', in
J. A. Moore (ed.), *Ideas in Modern Biology*, Natural History Press,
1965, pp. 176–89.

Cells as factories

In principle, the cell resembles a factory in requiring combustion
of fuel in order to provide energy for the performance of work.
However, there are some fundamental differences between man-
made factories and the cell with regard to the mechanism and
thermodynamic principles of energy transformation. In the
factory, the energy of combustion is used to do work by means of
heat engines. For example, combustion of coal can power a steam
engine, which can in turn perform mechanical work. But such
a heat engine can do work only if there is a temperature differential in the system. Such an energy transfer is not possible or is
highly improbable in the living cell, in which all the components
are essentially at the same temperature or, in other words,
isothermal. The cell is therefore unable to act as a heat engine,
and much more sophisticated devices for energy transfer have
had to be developed during the course of evolution of cellular
life. In brief, the cell carries out an essentially isothermal chemical
combustion of its foodstuffs and converts the chemical energy
released into another form of chemical energy, which is directly
used to drive the energy-requiring functions of the cell, again
under isothermal conditions. The cellular energy carrier is
adenosine triphosphate, which carries 'phosphate-bond energy'
from *exergonic* reactions such as respiration, which produce
energy, to the *endergonic* reactions such as active transport, biosynthesis and muscular contraction, which require energy (Figure
1). ATP is the 'charged' form of the energy-carrier system and
adenosine diphosphate (ADP) and adenosine monophosphate
(AMP) are the 'discharged' forms. It is the function of respiration to 'recharge' ADP and AMP to ATP by the process of

oxidative phosphorylation, in which oxidation–reduction energy is converted into phosphate-bond energy. The ATP so formed is discharged again when its energy is used to drive the endergonic functions. This cyclic process is a major activity. For example, in the kidney of the rat *in vivo* the turnover time of the terminal phosphate group of ATP is extremely low, of the order of fifty seconds, a fact which attests to the dynamic role of ATP in cellular energy transformations.

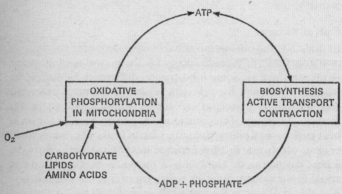

Figure 1 The energy cycle in aerobic cells

There is a second major difference between the factory and a cell. Many of the physical conditions under which energy is transformed in the operation of a factory are drastic and harsh – for example, high temperatures, high pressures, high voltages or currents, or extremes of pH. However, in the intact cell the energy transformations take place not only isothermally and rapidly, but at the rather mild temperature of 37 °C, in dilute aqueous systems of relatively low ionic strength, within narrow limits of pH and under very low electrical potential differences. Clearly the enzyme systems of the cell have been beautifully engineered by organic evolution for efficient and rapid energy conversions under mild isothermal conditions.

Finally there is a third difference between a cell and a factory, namely, the marvellous degree of miniaturization that organic evolution has produced in the cell. The machines involved in capturing and utilizing chemical energy in the cell are truly

molecular in action and in dimensions; they are therefore 'miniaturized' to the ultimate. These molecular machines are comprised of 'assemblies' of enzyme molecules; these enzyme assemblies are regularly arranged in the structure of the cell according to their function and purpose. The steric and dynamic organization of these structured multi-enzyme systems is under exquisite control, in part by genetic mechanisms, and in part by a complex network of feedback and other cybernetic systems, so that the molecular machinery for cellular energy conversion has self-adjusting and self-preserving properties.

The 'phosphate-bond energy' concept and cell structure

It is now over twenty years since Lipmann developed in broad terms the concept of the 'high-energy phosphate bond' in the energetic economy of the cell, following a number of important discoveries made in the 1930s regarding the role of phosphate and ATP in glycolysis and respiration. The picture Lipmann drew is essentially that which I have outlined in Figure 1. While these principles are still essentially correct, we are only just approaching satisfactory descriptions of the molecular biology of muscle action and of active transport, two of the most prominent energy-requiring functions of the cell. Today also the molecular mechanisms involved in the generation of ATP from ADP and phosphate by oxidative phosphorylation and photosynthetic phosphorylation are still essentially unknown in detail; they represent perhaps the greatest challenges in contemporary enzyme chemistry.

Perhaps the major difficulty in developing a detailed molecular basis for energy transfers in the cell may lie in the fact that much of the biochemistry of the past generation was relatively unconcerned with the chemical and physical significance of cell structure; biochemical mechanisms and theories were usually constructed in terms of ideal solutions and of homogeneous mixtures. Cell structure was regarded by many biochemists as a nuisance, standing in the way of purification of enzymes and substrates and their study in simple, idealized systems. However, today we recognize, as we can see from De Robertis's analysis of cell structure, that many if not all energy exchanges in the cell occur not in simple ideal solution, but rather in or on cellular surfaces or structures of great complexity. In fact, cell biologists

and biochemists alike have seen a transition from an earlier period when the cell membrane was regarded as an inert bag or container to hold the viable protoplasm in which the enzymatic reactions of metabolism were supposed to take place. Today it is the 'bag' and the other membranous systems of the cell that receive paramount attention. The more sophisticated and highly developed enzyme systems are actually associated with the membranes, whereas only the more primitive enzymatic processes appear to occur in the soluble hyaloplasm. Quite simply, full development of the mechanism of cellular energy transfer awaits a detailed molecular description of the ultrastructure of different portions of the cell such as the plasma membrane, the endoplasmic reticulum, the ribosome, the mitochondrion and the nucleus. However, we are now at least beginning to approach a point in the study of energy-transfer mechanisms in the cell where molecular structure may define functions and function may define molecular structure.

Adenosine triphosphate

Although ATP was first isolated over thirty years ago, and has been successfully synthesized in the laboratory, the usual textbook structural formula for ATP (Figure 2) does not begin to reveal its true three-dimensional structure nor all its molecular properties and features. It tells us nothing as to why this complex molecule, rather than some other derivative of pyrophosphate,

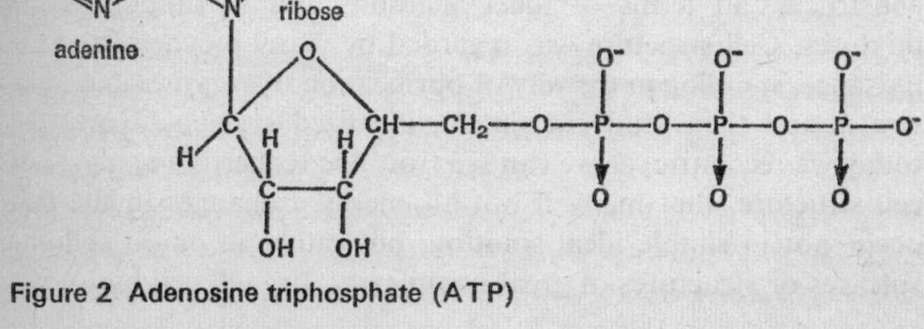

Figure 2 Adenosine triphosphate (ATP)

was selected during the course of organic evolution for its energy-carrying qualities. It also tells us little as to why it is a high-energy compound at all.

Actually, the true three-dimensional geometry of the ATP molecule is still unknown, despite the fact that we now have detailed information concerning the conformation of amino acids, peptides, one or two proteins such as myoglobin, and DNA. It is certain from existing X-ray diffraction and optical-rotation data, however, that the ATP molecule is not strung out in the extended form shown in Figure 2 but is bent or coiled; the terminal pyrophosphate group is believed to be in close proximity to the 6-amino group of the adenine ring. It is also most probable that a metal ion, Mg^{2+}, is normally a component in maintaining this structure. Such a conformation might help to explain why the adenine ring of ATP is specifically necessary for enzymatic reactions that involve chemical changes occurring in the phosphate groups at the other end of the molecule. But this is only a guess; we will see that other nucleoside triphosphates can also act as energy carriers.

Second, it is of some importance that ATP potentially may ionize in five separate steps and that each of the ionic species of ATP is probably stabilized as a resonance hybrid. Exact values for the ionization constants are not yet available. Determination of each of the microscopic ionization constants of such a polyelectrolyte is a formidably complex problem, since the separate steps influence each other. Thus ATP undergoes complex changes in ionic species with even slight variations of pH and ionic strength. That ATP is a polyelectrolyte has some important consequences for its function in the cell. One is that the relative proximity of its ionizing groups makes it possible for ATP to exist in the form of very stable chelates with various metal ions. Actually, very little free ATP exists as such in the intracellular medium; essentially all of the ATP in the cell is in the form of chelates with Mn^{2+}, Ca^{2+} and Mg^{2+}. The stability of these chelates decreases in the order named. Since the abundance of these ions in the cell is in the descending order Mg^{2+}, Ca^{2+}, Mn^{2+} (Table 1), it is evident that there is in the cell a complex distribution of the ionic and chelate species of the ATP molecule that is dependent on the pH and ionic strength of the medium, as

well as the momentary abundance of these divalent cations. Adenosine diphosphate, on the other hand, having one less ionizing group, is a less complex molecule. It can also form such chelates, but in general they are weaker than the ATP complexes.

Table 1 Stability Constants of Metal Complexes of ATP and ADP

$$K = \frac{[\text{Complex}]}{[\text{Metal}] [\text{Nucleotide}]}$$

Metal ion	ATP^{4-}	ADP^{3-}
K^+	11	6
Na^+	14	7
Mg^{2+}	10 000	1400
Ca^{2+}	5900	660
Mn^{2+}	56 000	8700

A second consequence of the complex ionization behavior of ATP is that the thermodynamic stabilization of each of the ionic species in an important factor in determining the magnitude of the free-energy changes when ATP undergoes hydrolysis during enzymatic energy- or phosphate-transferring reactions. In fact the so-called 'energy-rich' bonds of ATP and other phosphate compounds do not have exceptionally high bond energies or zero-point energies. Rather they are called energy-rich bonds because there is a relatively large decrease in free energy when they undergo hydrolysis. It is therefore the energy *difference* between ATP and ADP anions, at the specific pH of the system and in the presence of specific chelating metal ions, that endows ATP with its 'high-energy' characteristics. Actually, there is no particularly large free-energy difference between the un-ionized ATP molecule and un-ionized ADP and phosphoric acid.

Although earlier estimations had suggested a very high free-energy change of -50 kJ mol^{-1} for the reaction

$$\text{ATP} + H_2O \rightarrow \text{ADP} + \text{phosphate},$$

it is a feature of more recent and more penetrating investigation that the true free energy of hydrolysis is much lower, perhaps in the range of -24 to -29 kJ mol^{-1} under standard thermodynamic

conditions (i.e. molar concentrations of all components). However, although these new values have been widely taken as 'debunking' the high-energy nature of ATP, actually such an impression is perhaps misleading. Under intracellular conditions of pH and concentration, which are rather different from the standard conditions of the thermodynamicist, the actual value for the free energy of hydrolysis of ATP probably is substantially greater, and approximates -42 to -50 kJ mol^{-1}. Truly accurate measurements or calculations for this value under intracellular conditions must account for the different ionic species in the ATP–ADP–phosphate system in the presence of chelating divalent metal ions, the pH and the ambient concentrations of Mg^{2+}, Ca^{2+}, Mn^{2+}; the ΔF value for the hydrolysis thus may vary significantly depending on conditions.

Utilization of ATP by endergonic reactions

Let us now consider how the hydrolysis of ATP is utilized to 'drive', or more accurately, to 'pull' the three main energy-requiring functions of cells, namely, (1) the biosynthesis of large molecules from small precursors, (2) the mechano-chemical work of contraction and (3) the osmotic work involved in active transport.

In most energetically coupled chemical reactions, such as participate in the three processes just named, the transfer of energy from one molecule to another requires that the energy-yielding and the energy-absorbing reactions share a common intermediate. So far as is known, this principle of chemical thermodynamics underlies all reasonably well understood cases of energy transfer in cellular systems. The simplest case is given by the following equations for the familiar reaction of energy conservation in the glycolytic cycle:

3-phosphoglyceraldehyde $+ HPO_4^{2-} + DPN^+$

$$\overset{E_1}{\rightleftharpoons} \text{1, 3-diphosphoglycerate} + DPNH + H^+, \qquad \textbf{1}$$

1, 3-diphosphoglycerate $+ ADP \overset{E_2}{\rightleftharpoons}$ 3-phosphoglycerate $+ ATP$. **2**

Sum:

3-phosphoglyceraldehyde $+ HPO_4^{2-} + DPN^+ + ADP$

$\rightleftharpoons$ 3-phosphoglycerate $+ DPNH + H^+ + ATP$.

These two reactions bring about the conversion of energy of oxidation into 'phosphate-bond energy' of ATP. The principle of the common intermediate becomes clear when we break down the two reactions in a different way. The *exergonic* or energy-yielding process is the oxidation of the aldehyde to the acid:

3-phosphoglyceraldehyde $\rightarrow$ 3-phosphoglycerate $+ 2e$

$$\Delta F = -59 \text{ kJ.} \quad \mathbf{3}$$

The *endergonic* or energy-absorbing process is the formation of ATP from ADP and HPO_4^{2-}:

$$\mathbf{ADP + HPO_4^{2-} \rightarrow ATP + H_2O} \quad \Delta F = +42 \text{ kJ.} \qquad \mathbf{4}$$

In the overall process energy from reaction **3** has been transferred to reaction **4** by a reaction intermediate that is common to both processes. The common intermediate is 1, 3-diphosphoglycerate, which is a participant in reactions **1** and **2**. It is itself a high-energy compound since its hydrolysis to 3-phosphoglycerate $+$ HPO_4^{2-} proceeds with a large free-energy drop.

This principle of the common intermediate will be seen to operate in other types of energy transfer.

Utilization of ATP in biosynthesis of large molecules

In biosynthetic reactions, ATP acts as a donor of chemical energy for the creation of new chemical linkages that are characteristic of the structure of macromolecules such as glycogen, complex lipids, proteins and nucleic acids. In all these cases we shall see that the principle of the common intermediate between the exergonic and the endergonic reactions still holds, even though ATP itself may not be the immediate reactant that raises the chemical potential of a simple precursor molecule to a higher level and thus prepares it for the active biosynthesis. In Figure 3 we see that although ATP has been traditionally regarded as the primary energy source for the biosynthesis of these macromolecules, actually there exists in most cells a whole family of similar compounds, called nucleoside 5'-triphosphates (or NTPs), which differ only in the purine or pyrimidine base component. These include UTP, GTP and CTP. The discharged forms of the NTPs, namely UDP, CDP and GDP, can accept a phosphate group from ATP and become the immediate energy donors in the

synthesis of the new chemical bond. For example, uridine triphosphate, rather than adenosine triphosphate, is the immediate energy donor in the normal physiological synthesis of glycogen and certain other complex carbohydrates, a discovery made by Leloir and his colleagues. It is curious that the uridine nucleotides were selected during biochemical evolution specifically for polysaccharide synthesis.

In the case of the biosynthesis of complex lipids, cytidine tri-

Channeling of ATP Energy in Biosynthetic Reactions by Other NTPs

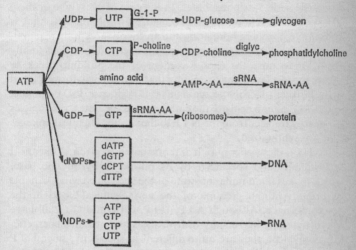

Figure 3 The utilization of phosphate-bond energy in biosynthetic reactions via different nucleoside 5'-triphosphates

phosphate is the specific energy donor for the activation of the nitrogenous bases such as choline and ethanoline that are incorporated into the structure of specific phosphatides, a fact first discovered by Kennedy. In the case of protein biosynthesis we find still another pattern. Amino acids are first 'activated', or raised to a higher energy content, by a reaction in which the amino acid is condensed with the adenosine monophosphate moiety of ATP to form an amino acyl adenylate bound to the activating enzyme, which is then enzymatically transferred to

soluble RNA. In the last step we see that the amino acid is transferred to the peptide chain by the action of still imperfectly understood 'adaptor enzymes', but here we see the reaction requires not ATP but guanosine triphosphate or GTP. Finally, we have the participation of four deoxyribonucleoside triphosphates in the case of DNA synthesis, specifically dATP, dCTP, dGTP, and dTTP.

In the case of RNA biosynthesis, specifically the biosynthesis of the so-called 'messenger' RNA, we see that the characteristic bases of RNA are incorporated starting from the corresponding ribonucleoside 5'-triphosphates. In the case of every one of these biosynthetic reactions we can recognize an intermediate that is common to an energy-donating and an energy-accepting reaction, which may be looked on as carrying the energy donated originally by ATP to raise the energy content of the building block or precursor to a level approximating that of the newly formed bond in the finished macromolecule. Thus UDP-glucose, CDP-choline, and AMP-amino acid are the common intermediates, the vehicles for energy transfer from the NTP to the specific chemical linkages that are synthesized.

Each of these energy-rich intermediates is given a chemical individuality and specificity by its purine or pyrimidine base, which makes it a unique substrate of the specific enzymes capable of bringing about creation of the new chemical bond in the macromolecule. Although ATP, UTP, CTP and GTP all have about the same free energy of hydrolysis and are thus energetically equivalent, they are quite different in specificity, since UTP is specific for carbohydrate biosynthesis, CTP for lipid synthesis, and so on. Ultimately, however, UDP, CDP and GDP must be recharged again at the expense of ATP, in reactions carried out by a series of enzymes called *nucleoside diphosphokinases*, which catalyse transfer of the terminal phosphate group in a reversible manner from ATP to the other nucleoside diphosphates, according to Table 2.

Although we now know that there is a whole family of NTPs operating in biosynthetic reactions, ATP is still the central member of the team. Only ATP can phosphorylate all the other NTPs. Furthermore only ADP can be rephosphorylated during oxidative phosphorylation. Nevertheless, the existence of four

NTPs permits diversification and control over biosynthetic reactions, particularly since the flow of energy among the different pathways may be dictated in part by the relative steady-state concentrations of the appropriate NTPs in the intracellular medium.

Table 2 Nucleoside Diphosphokinase Reactions

$UDP + ATP \rightleftharpoons ADP + UTP$

$CDP + ATP \rightleftharpoons ADP + CTP$

$GDP + ATP \rightleftharpoons ADP + GTP$

One other feature of these energy-transferring reactions involved in biosynthesis of macromolecules is noteworthy. We have seen in the cellular energy cycle that ATP is dephosphorylated to ADP and phosphate in driving the endergonic functions, and that ADP and phosphate are resynthesized to ATP during respiration. However, this statement is not entirely accurate. If we look at the equations for some of the biosynthetic reactions (Table 3), we find that in most cases it is not a terminal *orthophosphate* group that is transferred, or left behind, but a terminal *pyrophosphate* group. In the case of glycogen formation from glucose 1-phosphate and UTP, we find that the two terminal phosphate groups of UTP are left behind in the formation of UDP-glucose. This is a so-called 'pyrophosphate cleavage'. Similarly, in the case of preparation of choline for incorporation into lipids, the terminal pyrophosphate group of CTP is split out in the formation of cytidine diphosphocholine from cytosine triphosphate. In amino acid activation, we again find inorganic pyrophosphate as

Table 3 Pyrophosphate Cleavage in some Biosynthetic Reactions

---

1 $ATP + \text{amino acid} + sRNA \rightleftharpoons sRNA\text{-amino acid ester} + AMP + PP_i$

2 $ATP + \text{fatty acid} + CoA\text{-}SH \rightleftharpoons \text{fatty acyl-S-CoA} + AMP + PP_i$

3 $nNTP \rightleftharpoons (NMP)_n [RNA] + nPP_i$

4 $ndNTP \rightleftharpoons (dNMP)_n [DNA] + nPP_i$

5 $UTP + \text{glucose 1-phosphate} \rightarrow UDP\text{-glucose} + PP$

6 $CTP + \text{phosphorylcholine} \rightarrow CDP\text{-choline} + PP_i$

---

the end product, arising from the two terminal phosphate groups of ATP.

Studies of organic reaction mechanisms suggest that pyrophosphate is a better 'leaving' group than is inorganic orthophosphate. Pyrophosphate leaves the enzyme active site more readily because it is more highly charged than orthophosphate at pH 7 and because of the fact that it is more strongly stabilized by the formation of a resonance hybrid at the pH of the cell contents. But formation of inorganic pyrophosphate appears to be a rather wasteful process since the regeneration of ATP requires the reaction of ADP and orthophosphate. Pyrophosphate must first be hydrolysed to two molecules of inorganic orthophosphate before oxidative phosphorylation of ADP can occur (Table 4). Since the two terminal phosphate bonds of ATP are equal in energy content, the biosynthetic reactions result in the cleavage of *both* the high-energy phosphate bonds of ATP, one for the activation reaction itself, and the other for the cleavage of the inorganic pyrophosphate formed into orthophosphate. Thus these biosynthetic reactions are given an extra thermodynamic 'pull', a total of about 84 kJ. The nucleoside monophosphates formed are then rephosphorylated to the triphosphates in two stages (Tables 4 and 2).

Table 4 Fate of Pyrophosphate and Nucleoside Monophosphates

Pyrophosphatase	*Nucleoside monophosphokinases*
$PP_i + H_2O \rightarrow 2\,P_i$	$AMP + ATP \rightleftharpoons 2ADP$
	$CMP + ATP \rightleftharpoons CDP + ADP$
	$UMP + ATP \rightleftharpoons UDP + ADP$

Conversion of ATP energy into osmotic energy by active transport mechanisms

Active transport is defined as the movement of a substance in the direction of an increasing thermodynamic activity or concentration, usually across a membrane or across a barrier of cells. Such a movement is not spontaneous and can take place only if it is coupled to some other process capable of furnishing energy. While the theoretical thermodynamic treatment of the conversion of chemical into osmotic energy is simpler than that of mechanochemical coupling, it is on the other hand extremely difficult to

apply such thermodynamic considerations rigorously to transport mechanisms under cellular conditions. For this reason, active transport processes are experimentally recognized by simple metabolic tests rather than by thermodynamic criteria. If an *apparent* transport process is inhibited by a substance capable of blocking the formation or the utilization of ATP it is often regarded as a case of 'active' transport. Thus, poisoning of respiration by cyanide or dinitrophenol usually halts active transport processes.

All cells are capable of maintaining their internal ionic and solute environment relatively constant in the face of wide fluctuations in the external surrounding medium, by active transport mechanisms. This type of active transport has been termed *homocellular active transport* and is presumably a reflection of the biochemical adaptation of cells to changes in their external environment during the course of organic evolution. In the tissues of higher animals there exists in addition to homocellular active transport so-called *transcellular active transport*, in which the capabilities of the cell membrane have been exploited to achieve a directionality of transport so that there is a net movement of a substance across a cell barrier. Such transcellular transport is characteristic of epithelial cells, such as renal tubule cells, which are capable of secreting urine very different in composition from the blood from which it is derived. Still another biological adaptation of active transport mechanism is represented by the action potentials of nerve fibers, which are believed to be brought about by rapid, active transport of Na^+ and K^+ across the axon membrane.

Recent research is revealing some inklings, at both the conceptual and experimental levels, of the molecular mechanisms of active transport. At the conceptual level, the type of mechanism postulated by Mitchell has attracted much attention and seems probable to be correct in principle. In brief, he has proposed that active transport of H^+ and OH^-, as well as other substances, may be brought about by enzymes located in membranes in such a way that their active sites are asymmetrically arranged with respect to the plane of the membrane, so that the reactants or products must approach or leave the active site from either one or the other side of the membrane (Figure 4). Specifically, the

hydrolysis of a molecule of ATP by a membrane-bound ATPase, a process that requires a molecule of H_2O, may occur in such a manner that the H^+ is extracted from the aqueous compartment at one side of the membrane and the OH^- ion from the other side of the membrane, to form phosphate and ADP. In this way a pH gradient is created across the membrane by the hydrolysis of ATP. Since the membrane is relatively impermeable to H^+ and OH^- ions, it is possible that rather large pH gradients may be established in this manner. Mitchell points out that this is a

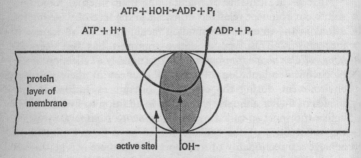

pH gradient produced by asymmetric arrangement of ATPase active site

Figure 4 Schematic representation of an anisotropic ATPase in a membrane. The active site of the enzyme is so located as to utilize H^+ ions only from one side of the membrane and OH^- ions only from the other side of the membrane in carrying out the formal hydrolysis of ATP to ADP and phosphate

vectorial chemical reaction, which has directionality. If the ATPase existed in ideal aqueous solution it would split ATP, but would not perform osmotic work; such a reaction is a *scalar* or non-directional reaction. Katchalsky has provided some theoretical thermodynamic principles that underlie vectorial processes in membrane active transport; these are most illuminating.

Recently a striking example of an asymmetric ATPase in a membrane has been described by Skou and subsequently by a number of other investigators. Skou found that the ATPase activity of crab nerve membrane required the presence of both Na^+ and K^+ in the medium for maximum rates. He found further

that the glycoside *ouabain* inhibited the stimulation of the ATPase activity by Na^+ and K^+; this substance has long been known to inhibit active transport of Na^+ and K^+. Skou postulated that this ATPase activity of crab nerve was responsible for the active transport of Na^+ and K^+.

The essential correctness of this idea was quickly established by Post, by Whittam and by Glynn in studies of erythrocyte membrane ATPase. Whittam in particular showed that ATP was split at maximal rates by erythrocytes when the external surrounding medium was rich in K^+ and the internal medium rich in Na^+; as an end result of the ATP hydrolysis, K^+ moved into the erythrocyte and the Na^+ moved out. Both ATPase activity and the ion movements could be abolished by ouabain. The ingenious experiments clearly established the directionality of the erythrocyte membrane ATPase to bring about active transport of Na^+ and K^+.

In principle these ion movements may be considered to take place by elaboration of the Mitchell concept. It seems possible that the enzyme is active only if K^+ is provided at one portion of the enzyme active site, from one side of the membrane, for one stage of the overall hydrolysis of ATP, and that Na^+ is required in a second stage of the hydrolysis but must be supplied from the other side of the membrane, each ion being discharged at the opposite side. Alternatively, Whittam has suggested the possibility that the ATPase molecule in the membrane may actually rotate during the hydrolysis of ATP so as to cause delivery of the ions on opposite sides. In any case this development, as well as recent research on active transport in mitochondria, may give us a new picture of the molecular biology of active transport.

3 P. Echlin

Origins of Photosynthesis

P. Echlin, 'Origins of photosynthesis', *Science Journal*, vol. 2,
1966, no. 4, pp. 42–7.

Fossils, far older than any previously known, have recently been
found in rocks of Precambrian age along the northern shore of
Lake Superior in North America. Radioactive dating techniques
suggest they are about 2000 million years old and they can be
seen microscopically to consist of minute spheres and filaments.
It has been proposed that these are the remains of organisms
which may be related to the modern blue-green algae – primitive
plant life forms found in soil, in fresh water and the sea whose
nearest familiar relatives are the green and red seaweeds. These
fossils may throw some light on the origins of the higher plants
and, in particular, on the way in which certain metabolic processes, used by the higher plants, have evolved.

The green colour of most modern plants is caused by the presence
in the individual cells of chlorophyll, a pigment used by the plant
as a catalyst in photosynthesis – the production of simple carbohydrates, such as sugars, from the inorganic compounds carbon
dioxide and water in the presence of light.

$$CO_2 + H_2O + light \xrightarrow{\text{chlorophyll}} (CH_2O) + O_2.$$

Photosynthesis provides an energy source for biological processes
by converting solar energy into stored chemical energy in organic
molecules. It is the ultimate energy source of all biological activity.
Man depends on photosynthesis for food and his store of fossil
fuels. Every year about 10^{11} tonnes of carbon dioxide are converted to carbohydrate by photosynthesis and, in terms of energy
conversion, this is many times more than all the energy man
currently produces from fossil fuels.

The chloroplast is the organelle in the cells of higher plants where photosynthetic energy conversion takes place and it can be considered as the photosynthetic counterpart of the mitochondrion – the organelle in the cell in which food energy is converted to chemical energy. Both organelles are more or less autonomous within the cell and both are major sites of energy conversion. The chloroplasts of higher plants and some algae are seen in the electron microscope to be enclosed by a double membrane and to contain dense areas or grana. These consist of closely stacked layers of discs each of which is made up of a pair of membranes (lamellae) each about 70 Å thick enclosing a space about 100 Å thick (1 Å is 10^{-8} cm). The grana lamellae are connected to less-dense lamellar structures, the stroma lamellae. This type of structure is also found in the blue-green algae and in certain of the photosynthetic bacteria. However, in these organisms the membranes of the photosynthetic apparatus are not contained within a discrete structure but ramify through the whole cell. Some other photosynthetic bacteria do not appear to have lamellae, the photosynthetic machinery being confined to spherical pigment-containing bodies dispersed around the periphery of the cell.

Although much is known about the photosynthetic machinery of a wide range of organisms, little is known with certainty about its earliest appearance in geological time or about the nature of the organisms in which it occurred. It is obvious from the vast deposits of fossil fuels that photosynthesis is an ancient biochemical mechanism. Indeed, Berkner and Marshall of the Graduate Research Centre in Dallas believe that the process probably started about 3000 million years ago, but did not reach its maximum until the development of a land flora about 400–450

Figure 1 (overleaf) Earth's history probably began about 5000 million years ago. The atmosphere was reducing and consisted largely of ammonia and methane. Oxygen began to accumulate in the atmosphere through dissociation by sunlight. The process is self-regulating and further increase would have only occurred through photosynthesis which began about 3000 million years ago. About 600 million years ago the oxygen concentration probably reached 1 per cent of the present atmospheric level and enabled a metabolism based on atmospheric oxygen to occur

Millions of Years Ago	Period	Solar Radiation Reaching Surface of Earth	Terrestrial Atmosphere
5000			H_2 N_2 NH_3 CH_4 H_2O CO_2 O_2
4000			
3000	Precambrian	visible light short-wave ultraviolet long-wave ultraviolet	
2000			
1000			
500	Cambrian Devonian Carboniferous		1% 10% 20%

Structural Fossils	Chemical Fossils	Cellular Metobolic Pathways
		↑ abiogenic production of organic compounds
		origin of life ?
		anaerobic heterotrophs
	Swaziland fig tree series chemical remains of bacterial systems ?	anaerobic heterotrophs using solar energy to make ATP
		anaerobic photoautotrophs making ATP
Bulowayen stromalites		microaerophillic photoautotrophs making more ATP
blue-green algae ?	phytane pristane $^{13}C/^{12}C$ ratio	aerobic photoautotrophs in local oxygen concentration
Soudan iron formations		
micro-organisms ?		
Gunflint chert micro-organisms	phytane pristane $^{13}C/^{12}C$ ratio	
blue-green algae ?		oxygen gradually becomes a more tolerable part of the atmosphere
bacteria ?		
first higher-plant cell ?		
metazoan cells ?		aerobic respiration
	Nonesuch shale oil	
higher algae marine evolution terrestrial plants terrestrial animals		organisms have metabolic machinery found in modern forms
		oxygen-evolving photosynthesis at maximum

million years ago. The indications are that, with slight variations during geological time, this maximum has been maintained up to modern times.

There is now good evidence that there was widespread biological activity as far back as half the usually accepted age of the earth (5000 million years). However, a recent investigation of haematite deposits dated at 3400 million years from the Fig Tree series in Swaziland – a sedimentary succession which also bears the 'first problematical vestiges of life' – and the reported finding in the same area by Barghoorn of Harvard University of the chemical remains of bacteria, may revise this figure.

Barghoorn and his associates have also discovered structurally preserved fossils in rocks – the Gunflint chert – along the northern shore of Lake Superior, which morphologically resemble modern blue-green algae. The rocks have been dated to about 2000 million years ago. The most abundant organism was a filamentous assemblage closely resembling *Oscillatoria* and *Lyngbya*, two modern blue-green algae. This was associated with spherical bodies which could have been unicellular blue-green algae or the endospores of filamentous forms. However, although most of the organisms found in this particular fossil bed closely resembled blue-green algae, some showed certain affinities with other groups of micro-organisms such as bacteria and fungi. In yet other instances there is believed to be no living counterpart.

Schopf and his colleagues, also at Harvard, have been investigating these fossil-bearing rocks and have found what appears to be the oldest definite occurrence of bacteria in the fossil record. Examination of the material in the electron microscope revealed rods and cocci which, in shape and size, resemble certain modern bacteria. Chemical studies of these Gunflint chert deposits indicate that the organic matter was probably produced by photosynthesis and that the fossil-bearing rocks also contain phytane and pristane – degradation products of chlorophyll which are found not only as derivatives of green-plant chlorophylls but also in bacteria and widely among animals.

Cloud and his associates in Minneapolis have recently made a similar study of rocks from the Soudan iron formation in north Minnesota which are estimated to be 2700 million years old. These rocks contain pyrite balls possessing microstructures

which may be of biological origin, although the morphological evidence is inconclusive. However, radioisotope studies and the finding of pristane and phytane are both consistent with a biological origin, though it is not certain whether they are endemic to the rocks.

The universe in which we live is believed to be composed largely of hydrogen and helium. This factor, together with terrestrial geochemical evidence, strongly suggests that at an early stage in its history the terrestrial atmosphere was strongly reducing and lacked oxygen. Oparin, Haldane, Calvin, Wald and many others have put forward ideas and speculations concerning the terrestrial development of life. Most workers in this field visualize the evolution of the simple cell through various stages. Colloidal aggregations with associated macromolecules gradually formed a limiting membrane, which in turn lead to the possibility of different concentrations or kinds of material being able to exist inside and outside this primitive cell. The first true organisms were probably able to live in the absence of oxygen but obtained their energy by partially oxidizing organic substances, such as some of the hydrocarbon originally produced without the help of living organisms. These forms gave rise to organisms in which energy from absorbed light added to the anaerobic metabolism and these in turn later developed photosynthetic systems. These primitive photosynthetic forms are believed to have given rise to organisms capable of the chemical breakdown of water into hydrogen and oxygen in the presence of light – photolysis – which consequently released molecular oxygen. Organisms which then used this oxygen to respire more efficiently established the metabolic foundations of modern organisms.

There have been several suggestions as to both the mode and time of origin of oxygen in the atmosphere. Berkner and Marshall have calculated that the atmosphere began to contain oxygen approximately 3000 million years ago. Free oxygen must be derived chiefly from the break up of water into hydrogen and oxygen, and can either occur by the photodissociation of water vapour by ultraviolet light or by the process which occurs during green-plant photosynthesis. Urey has shown that oxygen production by photodissociation becomes self-regulating;

the production of oxygen shades the water vapour from further ultraviolet radiation and thus only a very small amount of oxygen would build up in the atmosphere. It therefore seems certain that the increase of oxygen in the atmosphere could have occurred only as a result of photosynthesis.

Berkner and Marshall consider that there were two critical phases in the development of an oxygen-containing atmosphere. In the first phase, which is believed to have occurred at the beginning of the Palaeozoic era 600 million years ago, the oxygen level reached 1 per cent of present-day levels – permitting aerobic processes to occur which would provide organisms with many times more energy per molecule than the more primitive fermentation processes. But, due to high ultraviolet radiation levels, any living organisms would have been aquatic. It has been calculated that at least nine metres of water would be required to protect these primitive cells from the lethal effects of ultraviolet light. The second critical phase, which is thought to have occurred 420 million years ago at the beginning of the Cambrian period, was when the oxygen level reached 10 per cent of present atmospheric levels. This allowed the development of a land flora and fauna because, at such oxygen concentrations, sufficient ozone would have built up to shield the Earth from lethal radiation.

What organisms were living when photosynthesis began to develop? The fossil evidence points to the existence of the blue-green algae and the bacteria. The modern counterparts of these two groups of organisms – although containing complex biochemical machinery – are structurally simpler than higher plants and animals.

Except for a few anaerobes, chemicals known as porphyrins are found in all living cells. Some porphyrins, such as bacteriochlorophyll and chlorophylls, are strong absorbers of visible light, and play a vital part in photosynthesis. Geochemical evidence suggests that such compounds were also found in the earliest photosynthetic organisms. Indeed, all modern biochemical evidence indicates that the basic process in which sunlight is trapped to provide the energy for carbohydrate synthesis is the same in all organisms. What changed during the evolution of

photosynthesis was not the means by which solar energy was trapped but the metabolism involved in storing this energy in chemical form. The first photosynthetic organisms probably used gaseous hydrogen present in the atmosphere of hydrogenated elements to reduce carbon dioxide – also present in the atmosphere – and so form carbohydrates. These organisms were anaerobic but, through selection and mutation, evolved into forms which used not gaseous hydrogen but the hydrogen from water to synthesize carbohydrates. This distinction is important because only the second form dissociates water and hence evolves oxygen.

The first organisms to utilize energy from solar radiation were probably similar in many respects to modern photosynthetic bacteria. These micro-organisms are capable of carrying out a photosynthetic metabolism which differs from that of green plants and blue-green algae in that it does not use water as a reductant and hence does not evolve oxygen. Instead, it depends on the presence of extraneous oxidizable compounds – such as hydrogen sulphide – which are dehydrogenated with the simultaneous reduction of carbon dioxide.

The structures associated with photosynthesis in these bacteria are quite distinct from the lamellae seen in the blue-green algae. There is, however, a common membranous structure associated with all photosynthetic organisms – a membrane-bound space, the space being in the shape of a round vesicle in some of the photosynthetic bacteria and a closed pouch in all other photosynthetic organisms, including blue-green algae. The work of molecular biologists suggests that the photosynthetic activity is confined to small particles which themselves are located either on the membrane or within the pouch. These primary structures associated with photosynthesis appear in all the photosynthetic organisms examined so far, and it seems unlikely that such a structural unit evolved independently several times to give rise to the various photosynthetic organisms now found on the Earth. Although organisms similar to modern photosynthetic bacteria could have been the first photosynthetically active cells in the primeval atmosphere, they would make no contribution to the increase in the oxygen content of the air and, like their modern

counterparts, would probably only make a small contribution to the amount of organic matter.

The blue-green algae – the other group of organisms in the fossil flora of Precambrian rocks – were probably the first photosynthetic organisms to produce oxygen. The modern blue-green algae are similar in many respects to the present-day photosynthetic bacteria and both may have evolved from a common ancestor. They differ from photosynthetic bacteria, however, in that they produce oxygen during photosynthesis, and possess a somewhat different photosynthetic apparatus.

Hans Gaffron of Florida State University has shown that green-plant chloroplasts and possibly blue-green algae contain the components of a simple photosynthetic process, very similar to that of anaerobic photosynthetic bacteria, which cannot dissociate water into hydrogen and oxygen. The ability to produce oxygen may thus be thought of as a later evolutionary adaptation. Gaffron postulated that more than 2000 million years ago a new version of the already existing photochemical reaction appeared. The two systems combined, producing a system in which water could be photochemically decomposed and evolved.

Initially the presence of oxygen in the atmosphere may have been an objectionable by-product and was probably accepted by the ferrous ions thought to have been abundant in water at that time or quickly reacted with partially oxidized rocks. Gradually, however, free oxygen would have accumulated in the atmosphere on a large scale.

What chance would such organisms have had of surviving under the conditions prevalent on the primeval earth? The ability of certain blue-green algae (and bacteria) to fix atmospheric nitrogen would presumably have an adaptive advantage in organisms evolving in a nitrogenous reducing atmosphere. Taking the group as a whole, modern blue-green algae are more resistant to ultraviolet radiation than most other organisms. Extrapolating back to primitive forms, the possession of such a characteristic would endow a certain evolutionary advantage in an environment of increased ultraviolet radiation flux. Blue-green algal colonies of various kinds cover themselves with mineral particles – either by entrapping and binding extraneous grains or by secreting calcium

carbonate or silica. Such mineralization, which is also seen in some Precambrian species, may have served to shade the cells and colonies from all radiation wavelengths including ultraviolet.

The presence of a thick mucilaginous sheath would be an advantage in resisting desiccation and ultraviolet radiation; the ability of some forms to live at high temperatures, as in hot geothermal springs, further supports the idea that blue-green algae were among the first photosynthetic organisms on the earth.

Many evolutionary trees have been constructed in an attempt to establish the derivation of higher plants from primitive

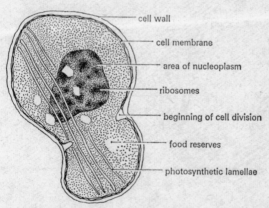

cell wall

cell membrane

area of nucleoplasm

ribosomes

beginning of cell division

food reserves

photosynthetic lamellae

Figure 2 Free-living blue-green alga *Merismopedia glauca* has a relatively thick cell wall in contrast to the thin walls found in the blue-green algae living in *Glaucocystis*. The lamellar membranes of the photosynthetic apparatus are not contained within a discrete structure but ramify through the whole cell. The nucleus is represented by an area of nucleoplasm

ancestors. Both bacteria and blue-green algae may have evolved to forms now extinct but which in turn may have given rise to higher plants. The unicellular bacteria may themselves have given rise to filamentous forms which, in turn, may have been the progenitors of the fungi. Similarly the unicellular blue-green algae may have developed various habits of growth and compartmentation of the specific sites within the cell, leading to the evolution of other algae and ultimately to the higher plants.

On both morphological and biochemical grounds the blue-green

algae share characteristics with the higher photosynthetic organisms. Dr Konrad Bloch of Harvard University has pointed out that the universal presence of the lipid a-linolenic acid in the photosynthetic apparatus of blue-green algae and higher plants, and its absence in photosynthetic bacteria, may mark the beginning of separate membrane-bound units of cell function. Similarly, galactosyl glycerides, the carriers for a-linolenate, are found only

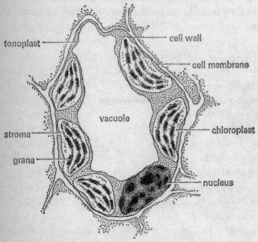

Figure 3 Cell of oat, *Avena sp.*, shows how the chloroplasts of a higher plant are enclosed within a discrete structure. Each chloroplast is surrounded by a double membrane and contains dense areas or grana which consist of closely stacked layers of discs, each made up of lamellae. The grana lamellae are connected to much less dense stroma lamellae

in photosynthetic organisms which evolve oxygen. It would appear that the ability to photolyse water may go hand in hand with the ability to synthesize lipids which, in turn, may have paved the way for the specialization and differentiation of cells. It is unlikely, therefore, that the photosynthetic bacteria gave rise to the higher organisms because of the relative morphological simplicity of their photosynthetic machinery, the absence of oxygen production during photosynthesis, and the difference in their photosynthetic pigments.

A totally different view is that the blue-green algae are the progenitors of the chloroplasts in higher plants. Such a hypothesis argues that the chloroplasts were not created *de novo* within the cell but that they were 'foreign bodies' somehow brought in from outside it. This idea is not entirely new and was first considered by two Russian biologists at the beginning of the century but not taken very seriously. However, recent advances in geology, biochemistry and the ultrastructure of cells have shown that the phenomenon of endosymiosis – in which one organism lives within the other and both are interdependent – may well be considered as a possible evolutionary step towards complex cell systems.

Going back to Precambrian times one may envisage a photosynthetic blue-green alga living in association with a non-photosynthetic organism. This may have been a fungal or bacterial-like structure which I will call a filamentous organism to avoid any confusion with later forms, as there is no direct evidence that such forms did exist at this time. Both organisms were in a reducing atmosphere and what simple intracellular structures they did possess are presumed to have been derived from the limiting cell membrane. The first stage in combining the two organisms is envisaged as a loose cellular association between the two, similar in many respects to the situation which exists in modern lichens where the cells of one organism live between the cells of the other. Passing through stages of increasing interdependence, the final stage would be the engulfment of the blue-green algae by the filamentous organisms, or invasion of the filaments by the blue-green algae, at which stage the host would be able to benefit fully from the products of photosynthesis. Thus, the structures which constituted the new organism followed a separate evolutionary course prior to their combining into the new cell.

The formation of a new organism from pre-existing organisms by means other than sexual and asexual reproduction is not so wildly fanciful as first appears. Professor M. R. Pollock of Edinburgh pursued this idea in a lecture to the British Association in September 1965. Workers in Oxford have succeeded in joining mouse cells and human tissue cells through artificial culture in the laboratory.

Nuclei fused during cell division and two sets of chromosomes

grew and multiplied within the same nuclear membrane. Protein synthesis has been achieved using components from several widely separated organisms, and the smallpox virus has been grown in bacteria. All these experiments demonstrate the interchangeability of biological systems.

There are living examples of such associations between blue-green algae and other organisms. The symbiotic associations are intracellular and quite distinct from the intercellular association that is seen in lichens. *Geosiphon* is a fungus (a Phycomycete) which contains within its cytoplasm intact living cells of a *Nostoc*-like blue-green alga. Blue-green algal-like organisms have been found growing symbiotically within the diatom *Rhopalodia*, in the colourless flagellates *Peliaina* and *Cyanaphora*, in the amoeboid *Paullinella* and in colourless algae. All these associations probably originated from a capacity to resist digestion by the host cell and usually result in a reduction in thickness of the blue-green algal cell wall. However, in order that it may remain within the host, it must retain the genetic mechanism ensuring replication into daughter cells.

The advantage, other than protection, gained by the blue-green alga is not entirely clear, but there may well be nutritive interrelations which await elucidation. The presence of these blue-green algae may profoundly modify the host's metabolism, for in *Peliaina* the alga produces plentiful amounts of reserve starch and individuals devoid of the algae often lack starch altogether. Thus, a substance can be formed by the dual organism which neither partner can produce by itself.

I have been studying for some time the green alga *Glaucocystis nostochinearum*. This does not contain chloroplasts but is classified as a 'green' alga because of the presence of starch as a food reserve, a cellulose cell wall and a curious polar thickening of the cell wall. It is able to live in the presence of light, water and simple mineral salts due to blue-green algae living within it, dividing as the host cell divides and acting as a chloroplast for the host. *Glaucocystis* itself was probably photosynthetic at one time but lost this ability only to regain it at some later date by the inclusion of blue-green algae into its cytoplasm. These have become sufficiently adapted to this mode of existence to lose the thick wall normally associated with the group and retain what appears to be

only a thin limiting membrane. It has not so far proved possible to grow the blue-green algae separately and, indeed, it may have lost its powers of independent growth. Experiments in my laboratory at the University of Cambridge have shown that the host alga can be cured of the blue-green algae if treated with penicillin and work is in progress to find out the nutritional requirements of such cells.

We can only guess, using our knowledge of modern organisms, how the initial intracellular combination actually occurred in primitive forms. The formation of a cell organelle by symbiosis implies that no organisms would exist which contained intermediate stages. It was an all or nothing process. Such a situation is reflected in the fossil record for there is a complete absence of such intermediate forms. There is a gap of about 2000 million years between the appearance of the first photosynthetic organisms and the appearance of green algae which may have contained intact chloroplasts.

There is also morphological evidence which supports the idea that blue-green algae may have been the forerunners of the higher plant chloroplast. A single spherical blue-green alga bears many resemblances to an isolated chloroplast of a higher plant. Both are multilamellate, possess chlorophyll-a and ribosomes for protein synthesis and, most important of all, possess fibrile of deoxyribonucleic acid (DNA) having a high degree of genetic individuality. Recent studies on the genetics and fine structure of chloroplasts have revealed the presence of DNA of the bacterial and blue-green algal type. This DNA is not organized into chromosomes and is not associated with basic protein as are the DNAs of higher plants and animals. Recent biochemical studies have confirmed the existence of chloroplast specific DNA. Messenger RNA complementary to chloroplast DNA, and a DNA-dependent RNA synthesis are further evidence for the comparative autonomy of the chloroplast within the cell.

There are also a few differences between blue-green algae and chloroplasts, including the absence of phycobilins from higher plants except in two algal groups, and the fact that the walls of the coccoid blue-green algae are much thicker than the bounding membrane of chloroplasts and contain diaminopimelic acid. There is good evidence that chloroplasts in modern plants do constrict and divide, as do blue-green algae, but there is also evidence that

chloroplasts may arise either from the nuclear membrane, or from pre-existing cell bodies in response to light. Although not entirely compatible with the blue-green algae theory of the origin of chloroplasts, these last two mechanisms may simply represent an evolutionary advance over the years.

The hypothesis that chloroplasts arose from blue-green algae remains unproven and, in any case, it is unlikely that it was the only evolutionary pathway. However, it is only by extrapolating back from what we know about the biology of modern micro-organisms, and using what little evidence there is about the palaeobiology of Precambrian microfossils, that we can make any postulates about the origins of photosynthesis.

4 G. Chedd

Recipe Complete for the Primeval Broth

G. Chedd, 'Recipe complete for the primeval broth', *New Scientist*, 28 March 1968, pp. 711–12.

To the cynical, the chemists and biochemists interested in the semi-philosophical problem of the origin of life often appear like the three witches in Macbeth, crouched over their bubbling cauldrons, hopefully conjuring nebulous images from the 'gruel thick and slab'. Certainly, there are grounds for this uncharitable suspicion. The experiments, begun in the early 1950s by men such as Urey, Miller, Abelson and Fox, have centred around the production of the organic chemicals essential to life in a laboratory reaction vessel containing only inorganic reagents. What gives these experiments their mystical aura is that everything about them can, in the nature of things, be only hypothetical.

The idea is to try to re-create the environment and conditions of the primitive Earth on the laboratory bench. Nobody knows for certain what things were like some four billion years ago, and, without the invention of a time machine, presumably nobody ever will. At the best, only intelligent guesses can be made. The atmosphere was almost certainly reducing, for instance, and only gradually gave way to the oxidizing atmosphere of today (probably itself largely the result of the evolution of life). Information gleaned from astronomy, astrophysics, chemistry, geology, meteoritic studies and biochemistry, has led to the majority belief that the commonest chemicals present in the primeval atmosphere were methane, ammonia, water and nitrogen. Deviations from the orthodox view include atmospheres consisting of carbon monoxide, nitrogen and hydrogen, and variations upon similar themes.

A similar mystery surrounds the source of energy necessary for the conversion of these simple inorganic compounds into complex organic molecules. Sunlight, lightning, vulcanism, ionizing

radiation – even the energy of meteoritic impacts – have all been suggested.

In general then, the contents of the researchers' cauldrons have included, at one time or another, almost all the possible variations of the probable constituents of the primordial atmosphere; and the witches 'fire' has included ultraviolet radiation, ionizing radiation, electric sparks and bunsen burners. Although not quite stooping to the addition of 'finger of birth-strangled babe' (biological contamination must, of course, be rigorously excluded), individual workers swear by other vital ingredients; for example, a handful of sand is reportedly a highly beneficial catalysing spice.

But, like the witches on the blasted heath, the people who have been working with these unpromising broths look like getting the last laugh. Over the past decade, just about every organic chemical found in living systems has been fished out of the sludgy mess which is usually the end product of their experiments: amino acids – which, linked together, constitute proteins; short chains of amino acids already linked together; protein-like molecules; the bases and sugars that go to make up the nucleic acids; fatty acids, and so on.

Just finding these chemicals proves, in itself, absolutely nothing of course; the step from small organic molecules to the large polymeric molecules of life, and the step from these large molecules to life itself, are both immense. But the discoveries do provide a firm foundation for further experiments and hypotheses. Perhaps more important still are the philosophical implications, for, although the problem of the origin of life has by no means been settled, the biochemical fishing in experimental primeval pools has shown that an answer is possible in terms of chemistry and physics, and without the invocation of any 'vital force'.

Despite the remarkable success of these experiments, however, two particular biochemical fish have consistently got away. Within the last few weeks, the landing of both of them has been reported in the literature.

Although very nearly all of the twenty-odd amino acids found in life have been detected in primitive-atmosphere experiments, one very important group has so far persistently resisted synthesis: the amino acids containing sulphur. This has been a

worrying omission, since sulphur-containing amino acids play a key role in the shaping and activity of proteins because of their ability to form bridges between and within protein chains.

Earlier this month a paper appeared from Steinman of Pennsylvania State University and Smith and Silver of Sir George Williams University, Montreal, reporting the isolation of minute quantities of the sulphur-containing amino acid methionine. Their reaction vessel contained ammonium thiocyanate (also found in volcanic gases as the result of the combination of carbon dioxide, ammonia and hydrogen sulphide), and the mix was irradiated with ultraviolet light.

Important though this discovery was, it is over-shadowed by the second report, by Hodgson and Ponnamperuma of the Exobiology Division of the National Aeronautics and Space Administration's Ames Research Center, Moffat Field, California, describing the isolation of a key chemical in the evolution of life – porphyrin, the major constituent of chlorophyll – as the result of firing a continuous electric arc through an atmosphere of methane, ammonia and water.

The experiment followed up an earlier, simplified, version, published last October by Hodgson and Baker, and carried out at the Research Council of Alberta, Edmonton. Porphyrin consists of a ring of four pyrrole molecules linked together. When a metal atom sits in the middle of the ring, the compounds become capable of absorbing the energy of sunlight and converting it into energy that can be used for chemical synthesis – in other words, they become photosynthetic. Chlorophyll itself consists of a porphyrin ring with a magnesium ion sitting in the centre.

Work carried out before the last war showed that porphyrins could be made by heating together pyrrole and formaldehyde – a very simple molecule – in a sealed tube for some thirty hours. Before Hodgson and Baker performed their experiment, they were aware that Ponnamperuma had some evidence that pyrrole is formed when methane and ammonia are sparked together in an electrical discharge. Formaldehyde is a common product of simulation experiments. Accordingly, Hodgson and Baker heated together pyrrole and formaldehyde, in the presence of crushed rocks (to provide the metal ions), in what was essentially a repetition of the pre-war experiments. Their reaction vessels

showed clear evidence that porphyrins, containing copper ions in the middle, had been made.

Encouraging though these experiments were, they did not, strictly speaking, show that metal porphyrins could have been formed in primitive earth environments. The evidence was only indirect: while it had been shown that chemicals capable of being made in simulation experiments could be heated together in isolation to form porphyrins, porphyrins themselves had still not been formed during the simulation process.

In fact, people have been looking very hard for porphyrins in their experimental vessels ever since simulation experiments began, with no luck. But as a result of Hodgson's and Baker's success, Hodgson joined Ponnamperuma from his laboratory in Alberta, and together they carried out yet another series of simulation procedures. Altogether, eighteen experiments involving sparking the methane–ammonia–water mix were performed. In each case, microgramme quantities of porphyrins were detected at the end of the experiments.

The experiments mark a major advance in the studies of pre-biological evolution. An essential step in this evolution is the build-up of polymeric molecules from the small organic molecules formed in the primitive oceans. Although several sources of energy for the synthesis of these small molecules are possible – as we have seen – the energy source with by far the greatest potential was sunlight. But only a very tiny fraction of the sunlight falling on the primitive oceans was of any use in chemical synthesis. The essential chemicals were therefore formed only very slowly, and were very, very few and far between. Primitive polymers forming from these building blocks were thus fairly soon faced with 'starvation' as they used up the pool of available chemicals.

The emergence of the porphyrins – now shown to have been possible under abiogenic conditions – changed all this with one fell swoop. Porphyrins are able to absorb the sun's radiation in the high-energy, visible region of the spectrum. Quite suddenly, a whole new source of energy was available for the synthesis of organic chemicals through the route of photosynthesis, and the first step had been taken along the road to the first primitive plants.

5 L. J. Rogers and T. W. Goodwin

When to Turn Green

L. J. Rogers and T. W. Goodwin, 'When to turn green', *New Scientist*, 25 January 1968, pp. 187–90.

The living organism is a complicated mechanism in which all the component chemical, physical and biological processes must be perfectly integrated and regulated. In higher animals, rapid inter-communication between organs and between organism and environment is possible because they possess a nervous system governed by the brain; longer-term regulation of living processes is achieved through hormones. In contrast, plants, lacking specialized cells for the rapid transmission of stimuli, are almost wholly dependent on slower-acting mechanisms based on hormones. But though the plant hormones such as auxin, gibberellin and abscisin have special roles in controlling growth, the plant must also be able to respond to signals from the environment.

One such light-induced response, which has interested scientists in Aberystwyth and Liverpool for some time, concerns the explosive formation of photosynthetic pigments triggered off by emergence of the germinating seedling from the soil. We now believe that this process (upon which fundamentally all higher forms of life are dependent) is brought about by a combination of segregation of enzymes within the cell and the impermeability of intracellular membranes to chemical precursors of the pigments. In simple terms, different compartments within the cell duplicate the enzyme systems for producing the chemical 'building blocks' needed to make pigments. The compartments do not, however, use these building blocks in the same way, nor do they normally cooperate with other compartments by supplying them with ready-made building blocks or even with the raw materials. But before we discuss this situation in detail we need to remind ourselves of some facts concerning the photosynthetic pigments; in particular their chemistry and their location within the cell.

Pigments found in higher plants include the green chlorophylls; the yellow, orange or red carotenoids; and the yellow, blue or red flavonoids. However, while the colours of flowers range from bright blue, mauve or purple to crimson, scarlet or bright red, those of leaves are, except in the case of variegated plants, some shade of green. This is because in leaves the green of the chlorophylls masks the yellow, orange or red of the carotenoids which are also present, whereas in flowers the flavonoids and the carotenoids predominate and chlorophylls are usually absent.

The chlorophylls and the carotenoids are found within the leaf cell in small lens-shaped bodies from four to six micrometres in diameter called chloroplasts. Their discovery is attributed to the seventeenth-century plant anatomist Nehemiah Grew, who spoke of air acting: '. . . the acid and sulphurous parts of plants for the production of their verdure; that is they all strike together into a green precipitate.'

It is in the chloroplasts that photosynthesis occurs, the primary act in this process being the capture of some of the sun's radiation by the pigments and its transduction into chemical energy in the form of adenosine triphosphate (ATP) and reduced nicotinamide adenine dinucleotide phosphate (NADPH). In this reaction, which is frequently called the light phase of photosynthesis, oxygen is liberated. The ATP and NADPH are essential for the second or dark phase of photosynthesis: this is the enzymatic conversion of carbon dioxide into carbohydrates and other organic compounds.

Chloroplasts are found mainly in the leaf, in its palisade and spongy mesophyll cells, though some do occur in the outer cells of the stem. Their characteristic structural feature is a complex fretwork system of lipoprotein permeating the space enclosed by the outer membrane. This lipoprotein fretwork is composed of small interconnected plates or lamellae which in some places are stacked to give structures called grana. The lamellae are suspended in an aqueous environment (the stroma) which contains soluble enzymes, carbohydrates, nucleic acids and other compounds. The chlorophylls and carotenoids are an integral part of the lamellae in which, therefore, the primary act of photosynthesis occurs. The dark reaction takes place in the stroma.

Chloroplasts usually form only when seeds are germinated in

the light. Seedlings germinated in the dark (etiolated) produce smaller organelles, the proplastids which, when the seedlings are illuminated, expand to the size of chloroplasts. Simultaneously internal structural rearrangements, additional synthesis of protein, and other changes, occur; these result in formation of the characteristic lamellar system of mature chloroplasts. The chloroplast pigments are also formed during this period.

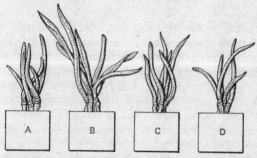

Figure 1 The effect of red light and far-red light on the growth of etiolated maize seedlings. A: a five-day-old seedling grown entirely in the dark; B: as A except exposed to red light for ten minutes after four days. If seedlings were exposed to far-red light for ten minutes following irradiation with red light (C) or were exposed only to far-red light for ten minutes instead of to red light (D), their growth was similar to that in A

If a seedling germinating in the dark is exposed to a short flash of red light (wavelength maximum, 6600 Å), and returned to darkness, then, during the next twenty-four hours, the proplastids swell to the size of chloroplasts but do not produce lamellae or synthesize chloroplast pigments. This effect can be nullified if the flash of red light is followed within about half an hour by a flash of red light from the far end of the visible spectrum (wavelength maximum, 7300 Å). The plant remembers only the last irradiation in the sequence. This reversal phenomenon is characteristic of photoresponses mediated by a substance called 'phytochrome'. Phytochrome is a blue chromoprotein in which the coloured component is a linear tetrapyrrole (closely related to the pigments we excrete in our bile). It absorbs light at 6600 Å and irradiation at this wavelength transforms it into an unstable form with an absorption maximum at 7300 Å. This is the

biologically active form of phytochrome. The unstable form reverts to the stable form slowly in the dark, and spontaneously when illuminated with the appropriate far-red light.

stable   unstable

$$\text{phytochrome} \xrightleftharpoons[\substack{\text{far-red light} \\ \text{or} \\ \text{slowly in the dark}}]{\text{red light}} \text{phytochrome}$$

What appears to be happening during the phytochrome-stimulated development of proplastids is, among other things, a synthesis of enzymes concerned with the formation of chlorophylls and carotenoids. When a phytochrome-stimulated seedling is illuminated, pigment synthesis 'takes off' almost immediately, while with seedlings not stimulated, there is a considerable delay before synthesis begins. During the lag period, the plastids are synthesizing enzymes they already possess in the dark if they have been 'treated' with phytochrome.

As well as plastid development, phytochrome appears to control many other photomorphogenic responses, such as germination of seeds, dormancy, growth of etiolated leaves (Figure 1), flowering, tuberization, and formation of flavonoids. How can one mechanism possibly account for all these changes? All that is known is that this control is not exerted at the genetic level; for example, the sleep movements of *Mimosa*, controlled by the state of phytochrome, occur within a few minutes of stimulation. One view currently held is that phytochrome controls membrane permeability and through action at this level can trigger off physiological processes. In contrast phototropism, the growth of a plant towards or away from the light, appears to be controlled by some as yet unknown yellow pigment capable of absorbing blue light. As a plant leaf ages, or as a young (green) flower (such as a buttercup) buds or unripe fruit (green tomato, say) matures, the lipoprotein fretwork of the chloroplasts may disintegrate and be replaced by yellow or red coloured globules or bundles of fibres. The chloroplast has become a chromoplast, a plastid without photosynthetic activity. It is the chromoplasts which contribute to the vivid colours of mature flowers (buttercup, for example) and fruits, such as the tomato. In the chromoplasts, the

chlorophylls have more or less disappeared and the yellow, orange or red colours of the carotenoids, which are still being actively formed, now predominate.

At the chemical level, control of pigment formation in leaves is primarily the problem of the control of terpenoid (isoprenoid) synthesis, since the chloroplast pigments are either wholly or partly terpenoid in nature. Terpenoids are derivatives of the 5-carbon compound isoprene (C_5H_8) and are formed in nature from isopentenyl pyrophosphate (IPP), the universal biological isoprene unit (Figure 2). The first specific terpenoid precursor, which is derived from three molecules of acetyl-coenzyme A (acetyl-CoA), is mevalonic acid (MVA) which is converted into IPP by two successive phosphorylations at the fifth carbon atom (C-5) followed by removal of carbon dioxide and water from the molecule.

Terpenoids in chloroplasts of green leaves include the carotenoids (mainly β-carotene (a carotene) and lutein (a xanthophyll)), the side chain of chlorophyll, plastoquinone, and some other compounds, and a small amount of sterol. Outside the chloroplasts both the sterols, which are involved in membrane formation, and the side chain of ubiquinone are terpenoids. The steps in formation of these important constituents of plant cells are common to the level of the compound farnesyl pyrophosphate (FPP) which has fifteen carbon atoms (Figure 2).

The necessity for control of terpenoid synthesis in the plant can be realized when the requirements of the plant at different stages in its life-history are considered. During germination, the seedling needs to form sterols from food stores in the seed for the manufacture of membranes and lamellae. However, when the seedling emerges from the soil and is illuminated, the chloroplasts develop rapidly and a priority at this time is the conversion of atmospheric carbon dioxide into the chloroplast pigments that are vital for photosynthesis. Although there is some continual renewal of existing sterol, a net sterol synthesis is a secondary requirement at this time. When the chloroplasts have matured, some of the products of photosynthesis are able to move out of the plastid and be converted into sterol necessary for further growth of the plant.

Thus we find that on illumination of an etiolated plant, there

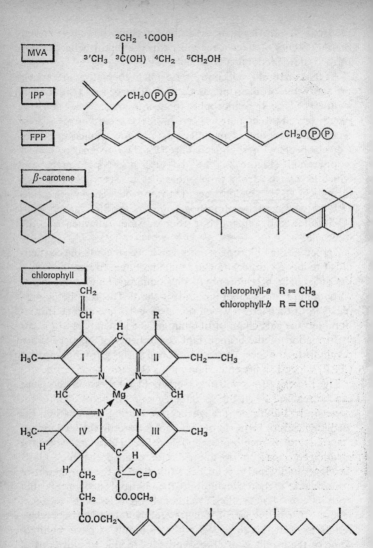

Figure 2 The chemical structures of some chloroplast pigments and their metabolic precursors. P = phosphate radical

is a rapid net synthesis of chloroplast pigments and other chloroplast terpenoids but not of sterols and ubiquinone. The synthesis of the chloroplast terpenoids parallels development of the chloroplast. Some as yet unknown self-regulatory mechanism comes into play in mature chloroplasts and prevents the formation of excessive amounts of chloroplast pigments.

All our evidence so far leads us to believe that this direction of terpenoid formation at different stages of plant growth is achieved through a combination of enzyme segregation within the cell and impermeability of the chloroplast membrane to terpenoids and their immediate precursors (Figure 3). It is suggested that enzymes to convert acetyl-CoA through IPP, geranyl pyrophosphate (GPP), farnesyl pyrophosphate (FPP), and geranyl-geranyl pyrophosphate (GGPP) to a C_{50}-pyrophosphate exist both inside and outside the chloroplast. However, only the chloroplast can convert carbon dioxide into acetyl-CoA, form carotenoids and the side chain of chlorophylls from GGPP, and form the side chain of plastoquinone from the C_{45}-pyrophosphate. Conversely, only outside the chloroplast can FPP be converted into squalene, and the side chain of ubiquinone be formed from the C_{50}-pyrophosphate. For this system to be effective, the membranes surrounding the chloroplasts must be almost impermeable to intermediates in the biosynthetic pathway. For example, it is inefficient to form MVA for chlorophyll and carotenoid synthesis in the developing chloroplasts if it leaks out to be wasted in the formation of unnecessary sterols.

Evidence in favour of the scheme has come from studying the incorporation into terpenoids of molecules labelled with radioactive carbon atoms and from enzymatic studies. It can be shown, for example, that the radioactive carbon atom from $^{14}CO_2$ is actively incorporated by illuminated seedlings into β-carotene and the side chains of the chlorophylls and plastoquinone, whereas little or no radioactivity is detectable in sterols and ubiquinone. However, if (2-^{14}C)MVA (i.e. mevalonic acid labelled with radioactive carbon in position 2) is used, the chloroplast terpenoids are not significantly labelled (since the chloroplast membrane is relatively impermeable to MVA) but the sterols and side chain of ubiquinone are highly radioactive.

Enzymatic studies have been facilitated by the use of methods

recently devised whereby virtually intact chloroplasts, free from other cellular organelles such as nuclei or mitochondria, can be isolated from leaves. One of the methods we use involves freeze-drying the leaves, homogenizing them in a carbon tetrachloride–hexane mixture, and separating the chloroplasts by centrifugation in a density gradient composed of mixtures of varying proportions

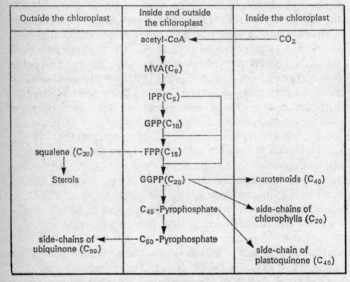

Figure 3 A simplified scheme indicating which reactions in terpenoid synthesis occur both inside and outside the chloroplast (central column) and which occur in only one of these sites (side columns)

of the two organic solvents. Chloroplasts prepared in this way contain very active enzymes including, we have discovered, mevalonate kinase, which catalyses the first phosphorylation of MVA in its conversion to IPP. Whereas the optimum pH of the chloroplast mevalonate kinase is 7·5, that of the same enzyme from outside the plastid is at pH 5·5 – strong evidence for the validity of the scheme shown in Figure 3. As we can predict from the scheme, intact chloroplasts will not metabolize MVA to any great extent but will do so readily after being broken open by

ultrasonic waves or osmotic shock. Evidence from other laboratories suggests that the formation of the porphyrin part of the chlorophyll structure is similarly controlled and such regulatory mechanisms may be a common feature of biosynthetic pathways in plant cells.

It will be some time yet before our understanding of the control of pigment synthesis in leaves is complete. However, the problem is an important one, since all higher forms of life are directly or indirectly dependent for their food and energy on the trapping of the sun's energy by the leaf pigments in the chloroplasts. Even when agricultural productivity is at maximum efficiency, the rate of food production, limited by the photosynthetic efficiency of the crops, may be insufficient for an expanding world population. If this occurs, knowledge of how to adapt or modify the process of pigment formation in chloroplasts may be of vital importance for human survival.

6 J. A. V. Butler

How Genes are Controlled

J. A. V. Butler, 'How genes are controlled', *Science Journal*, vol. 2, 1966, no. 3, pp. 41–5.

The brilliant work of molecular biologists over the past ten years has shed light on one of the most fundamental characteristics of life: the way in which hereditary information is passed on from generation to generation. The story is as yet by no means fully told, but the basic principles involved are well understood. It has even been possible to work out the chemical sequences in the genes which are needed to specify particular genetic instructions. Ultimately, these instructions concern the synthesis of proteins in the cell, chemicals which are used by life as enzymes or catalysts in the production of cell material.

However, many questions about how these genetic processes are themselves controlled remain unanswered. For instance, not all the genes are active all the time. Animals and higher plants are composed of numerous different types of cell and in each a different set of proteins is required. As every cell contains a full set of genes, some mechanism must exist for activating only those genes required by one particular kind of cell. Furthermore, some bacteria synthesize chemicals only when they are called for: here again, some gene-activation mechanism must be in operation.

In this article I shall discuss the various ways in which gene operation may be controlled, a process in which chemicals known as histones – the subject of this article – may be particularly important. Before doing so, however, I will clarify the basic means by which genetic information is carried in the cell.

The primary task of the genes is to carry information necessary for the synthesis of a large number of highly specific proteins from amino acid subunits. This information is carried in the form of a code based on the precise sequence, along the DNA

molecule, of four nucleotide units – adenine (A), thymine (T), cytosine (C) and guanine (G). The essential characteristic of these nucleotides is that adenine is complementary to thymine and cytosine to guanine and, because of this complementarity, DNA is able to replicate itself. A complete DNA molecule consists of two strands – a double helix – one being the exact complement of the other. Wherever adenine and cytosine are found in one fibre, thymine and guanine respectively are found in the other. This

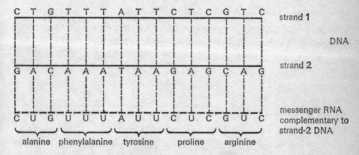

Figure 1 The genetic code for a group of amino acids as it appears in DNA and the corresponding messenger RNA. The DNA nucleotides cytosine (C) and guanine (G) are complementary to each other, as are thymine (T) and adenine (A). In messenger RNA thymine is replaced by uracil (U). The coding is ternary, three nucleotides specifying one amino acid

means that the two strands are related like a lock and key and that, if they are separated, each is capable of making its complement so as to reconstruct the whole DNA molecule.

The genetic code itself acts through an intermediary substance, ribonucleic acid (RNA), which is similar to DNA. It differs in that the sugar deoxyribose – found in DNA – is replaced by ribose in RNA and the thymine nucleotide is replaced throughout by another containing uracil which is also complementary to adenine. Fibres of RNA are formed on the DNA, as on a template, the order of the units in the RNA being guided by one of the strands of DNA. In this way the DNA code is transferred to RNA which is then used in various ways. At present, the order of all the nucleotides in a DNA gene has only been established in one instance. It happens that the RNA formed by this particular

DNA is of a type called 'transfer RNA' because it is used to transfer an amino acid – in this case alanine – from the free state into a state of combination in a protein.

We can now see in outline how the code carried by the DNA is used to guide the assembly of a protein. First, a strand of a type of RNA – known as 'messenger RNA' – is formed on the DNA. This strand becomes detached from the DNA or, probably, is pulled off by a subcellular particle, a ribosome. The ribo-

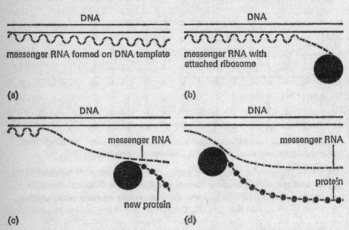

(a) messenger RNA formed on DNA template

(b) messenger RNA with attached ribosome

(c) messenger RNA / new protein

(d) messenger RNA / protein

Figure 2 The synthesis of a protein is carried out by a ribosome using the code carried by a messenger RNA. Transfer RNA does the actual attachment of the amino acid to the growing protein chain.
Messenger RNA is formed on a DNA template (a). A ribosome then attaches itself to one end of the messenger RNA and begins to pull it away from the DNA (b). It then moves along the messenger RNA 'reading' the code (c) and forming a protein chain (d) whose amino acids are in order set by code

some then moves along the fibre of messenger RNA and 'reads' the code. As it does so amino acids are added to the growing protein chain by the transfer RNA in accordance with the instructions given by the code.

It has been established that the code is almost certainly a ternary one, three units of the messenger RNA being required to specify each amino acid. The code sequences for most amino acids have also been established, though not with complete certainty:

CUG is the code for alanine, UUU for phenylalanine, AUU for tyrosine and CUC for proline.

It is a cardinal principle of cell biology that the full complement of genes is present in all, or nearly all, of the cells of the organisms. This can be demonstrated with plants and primitive animals in which a small part can often regenerate the whole. It can be inferred in other cases from the fact that, when allowance is made in some cases for duplication of the genetic units in each cell, nearly all the cells of an organism contain the same amount of DNA.

If every cell carries all the genes which are required in the many cell types which make up the organism, the amount of DNA will be much larger than in unicellular organisms such as bacteria. In fact, the amount of DNA per cell in animals is several hundred or even a thousand times as much as in typical bacterial cells. It follows from this that only a part, perhaps a small part, of the genetic code is actually being used in each kind of cell and much of it must be permanently non-functional.

Furthermore, bacteria, although they are not differentiated into different types of cells, have primitive gene-control systems. They are capable of making a large variety of enzymes, which perform chemical operations on food materials, but they only produce these enzymes when they are actually required. For example, when presented with an unfamiliar type of sugar, such as galactose instead of glucose, the bacterium *Escherichia coli* is at first unable to utilize it but, after being in contact with the substance during an 'induction period', the necessary enzyme is produced.

This phenomenon has been much studied in recent years and it has been established that it only occurs when the organism contains the gene for the new enzyme but which remains 'turned off' or inactive until required. When a new substance is offered, something happens to 'turn on' or activate the appropriate gene, which will then produce the messenger RNA for the required enzyme.

In 1963 a theory to account for this was proposed by Monod and Jacob at the Pasteur Institute in Paris. In extensive studies on the genetics of micro-organisms (especially *E. coli*) they

recognized three kinds of genes which they called 'structural', 'operator' and 'regulator' genes respectively. The messenger RNAs are actually made by the structural genes, which are in groups controlled by an operator gene. The latter is turned on or off by a regulator gene.

Jacob and Monod suggested that the control of an operator

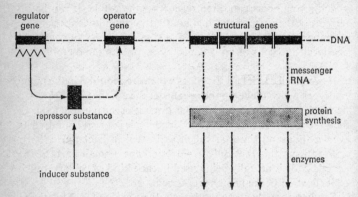

Figure 3 The method of control of protein synthesis as postulated by Jacob and Manod in 1963. They suggested three types of genes: structural genes carrying the codes for the synthesis of enzymes; operator genes capable of controlling the activity of a number of structural genes, and regulator genes in turn controlling the operator genes. It was suggested that a 'repressor' substance continuously produced by the regulator genes keeps the operator gene turned off so long as it reaches it. However, when inducer substances are present they combine with the repressor substance and prevent it reaching the operator gene or, if it does reach the gene, modify it in some way so that repression no longer occurs. Upon release from inhibition in this way, the operator gene activates its structural genes which are then able to produce messenger RNAs for the synthesis of needed enzymes

gene was brought about by a 'repressor' substance – produced continuously by the regulator gene – which keeps the operator gene turned off so long as it reaches it. It was suggested that when inducer substances (substances on which an enzyme can act) are present they combine with the repressor substance and prevent it reaching the operator gene. If, however, it does reach the gene, the inducer substance may modify it in some way so that repression no longer occurs. Released from inhibition, the operator

gene activates its structural genes which are then able to produce messenger RNAs for the appropriate enzymes.

In some instances, it has been found that the operation of a whole series of genes is controlled by a single product. Thus, the synthesis of a whole series of enzymes concerned with the conversion of ornithine to arginine is controlled by the presence of the final product. This prevents the production of more arginine than is required and is an example of feedback control.

In these circumstances it has to be supposed that the endproduct of a series of reactions can act as an inhibitor of the operator gene concerned, thus providing another method of gene control. The possibility also exists that the same substance under different circumstances can be either an activator or an inhibitor.

It has been found in recent years that proteins may behave in a number of different ways. Some enzymes, for example, have a number of reactive sites which are active under different circumstances. Thus, combination of one of these enzymes with a small molecule, which might be produced by a different enzyme, will change the nature of the original enzyme activity. Similarly, a protein could act as a gene activator or repressor depending on whether or not a small molecule is attached. Such proteins have been called 'allosteric' proteins by Jacob and Monod.

Yet another means of control lies in the existence of isoenzymes – enzymes with similar activities. It has been suggested that each iso-enzyme responds to a different kind of control so that multiple controls of enzyme activity are possible.

Thus there are a number of possible ways by which gene activity can be controlled in bacteria although some remain rather hypothetical. In the chromosomes of certain bacteria it has been established that the genes are arranged in a circle. Replication of the genes is brought about by an enzyme (DNA polymerase) which moves round the circle beginning at an attachment site and completing one replication cycle before another can take place (see Figure 4). Interestingly, Berg, Kornberg and others have shown that the enzyme RNA polymerase, which forms the messenger RNA on the active genes, can become attached to, and begin to act on, a number of sites on the cycle.

An attractive hypothesis would be to suppose that the

regulatory genes provide points of attachment for the RNA polymerase and provide a kind of gate through which the enzyme (polymerase) can reach the operator genes. The gate will be open or closed according to the state of the repressor protein. Activating substances will cause the gate to open, while the presence of inhibitory substances will close the gate (see Figure 5).

Thus, the various kinds of control can be explained; but so far little has been done to isolate and identify the substances postulated and to demonstrate that they do, in fact, operate in this way.

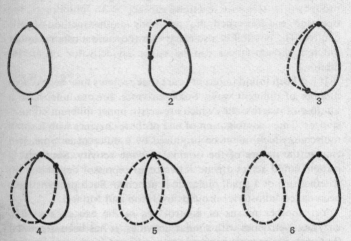

Figure 4 Replication of a circular bacterial chromosome is brought about by an enzyme – DNA polymerase – which moves around the circle beginning at an attachment site. One replication cycle, illustrated above in six phases, must be completed before another can take place

In complex organisms, however, much more complicated mechanisms of gene control are required. In any type of cell some of the genes must be active while others must be permanently inactive and there may also be controls of the type met with in bacteria where activity is induced by the presence of a substrate. Feedback controls may also be present to ensure that enzymes are only formed as required. In addition there are hormonal controls which are peculiar to complex organisms. It has been known for a long time that certain types of cell processes are

initiated, or accelerated, by hormones – substances produced in endocrine glands which stimulate the cells or other organs to secrete proteins or to act in other ways. It has been found recently by Tata, Korner and Kirby that specific hormones can stimulate the formation of messenger RNA or, in some cases, change its characteristics in the organs on which they act.

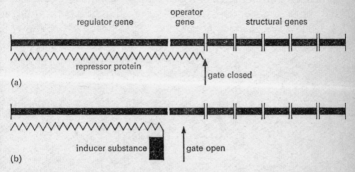

Figure 5 Regulator genes could provide points of attachment for RNA polymerase and provide a 'gate' through which the enzyme (polymerase) can reach the operator genes. The presence of repressor protein (a) would close the gate while inducer substances (b) possibly becoming attached to the repressor protein would cause the gate to open allowing access

How are all these effects to be fitted into the picture? Let me state at the outset that our knowledge is extremely rudimentary. In the chromosomes of both plants and animals it was found, a century ago, that the cell nucleus contains a special kind of protein called a 'histone'. These proteins are basic in character, containing a high proportion of the 'basic' amino acids, lysine and arginine. Because of this, they are capable of combining with acidic phosphate groups of DNA, forming spirals round the DNA fibre (see Figure 6).

These substances were discovered by the Swiss chemist Friedrich Miescher in 1868 at the same time as he discovered the substance he called 'nuclein' – now called DNA. However, little was found out about them until after 1950, when the subject was reopened by Stedman and Stedman in Edinburgh who showed that histones were a complex group of proteins and began the process

of separating them into their components. They suggested that the histones acted as gene inhibitors preventing the functioning of inactive genes. This has been confirmed recently, to some extent, by Huang and Bonner, Allfrey and Mirsky, Barr and myself and others who have shown in *in vitro* experiments that histones inhibit the formation of messenger RNA on DNA templates.

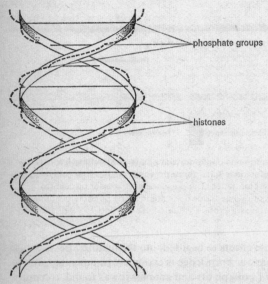

phosphate groups

histones

Figure 6 Histones — basic proteins containing a high proportion of the amino acids lysine and arginine — combine with the acid phosphate groups of DNA and form spirals around the DNA fibre. It has been suggested that they act as gene inhibitors preventing the functioning of inactive genes, and it has been shown that they inhibit the formation of messenger RNA on DNA templates

There remain many difficulties to this theory which suggests that active genes are free from histones and inactive genes are combined with histone. Thus, it is doubtful if there are a sufficient number of distinct histones to account for the different situations. Only about twelve distinct histones have been distinguished analytically with certainty although there are possibly a number

of others present in small amounts. It is conceivable that each of the major histone fractions contains a large number of distinct proteins of a similar character but differing slightly in their amino acid sequences; however, available evidence, though not very conclusive, does not support this.

Another argument which has been put forward is that if different sets of genes are active in different tissues we should expect to find different groups of histones missing in each type of cell. Although exact quantitative analyses of the relative amounts of histones present are not yet available, it is found that qualitatively the histones of different tissues, and even in different animals, are remarkably alike.

This is true of many mammals and also fish; recently Dr J. Palau of the Chester Beatty Research Institute found that trout liver contained the same histones in approximately the same proportion as, for example, the thymus gland of the calf. It is a remarkable thing that this group of histones has continued to occur in substantially the same form and similar amounts in the millions of generations which have passed since the trout and the calf had a common ancestor. The different kinds of histones obviously have very important functions.

If each gene or group of genes is to be associated with a particular type of histone, there must be some means by which a histone can recognize its proper genes. Also, of course, there must be mechanisms for removing it from those cells in which the gene is active. Experimentally there do not seem to be any special interactions between particular histones and different parts of the DNA.

The simplest solution to the problem of associating particular histones with specific genes would be to suppose that they are actually formed *in situ* or, at least, very close to where they are found. This is probably quite reasonable as ribosomes are found closely associated with the chromosomes and it is possible that protein could be synthesized even before the messenger RNA gets clear of the genes. A very interesting suggestion has been made recently, by Patel and Wang, in that DNA is itself capable of guiding the synthesis of protein directly, without the intermediate formation of the messenger RNA. If this is correct, it

would mean that every gene should be able to make its own protein *in situ*.

Nevertheless, it is doubtful if there are a sufficient variety of histones to provide all the kinds of gene control required in complex organisms. Moreover, although regulatory proteins are undoubtedly present in the bacterial chromosomes, they are not of the histone type. Little is known of these proteins at present, but they are probably similar to the so-called 'residual proteins' (non-histones) which are also found in the chromosomes of complex organisms. It is possible, and indeed likely, that these are also concerned with gene control.

It has recently been suggested by Hotta and Bassel that in animal chromosomes, as in bacteria, the DNA is in the form of circular loops joined together by a backbone, which may also consist of a segment of DNA. Loops of this kind are seen in 'lampbrush' chromosomes (see Figure 7). If this is the case the backbone might well contain the regulator genes while the operator and structural genes are present on the loops. The regulator sections might contain the 'residual' proteins which are sensitive to various substances capable of stimulating gene action, such as inducers and hormones. Some evidence in favour of such a view has been obtained by Talwar, Segal and their colleagues who isolated a protein from rat uterus which inhibited RNA synthesis by itself but which, when combined with the female sex hormone oestrogen, stimulated the synthesis of messenger RNA.

The mechanisms controlling gene action in multicellular organisms are of great complexity and have a multiple character. Every gene may be controlled in a number of ways, and there may be many genes whose sole function is to control other genes. In addition to the processes controlling particular biochemical reactions, there must also be present overall controls such as those which regulate cell division and keep the growth of different

---

Figure 7 (a) Model for part of a mammalian chromosome which shows how the various genes may be arranged. It is suggested that the 'backbone' consists of a DNA fibre to which are attached regulatory proteins and that the loops consist of operator and structural genes. (b) A segment of a lampbrush chromosome taken from an oocyte of the newt Triturus. Loops consist of RNA fibres round a delicate DNA fibril

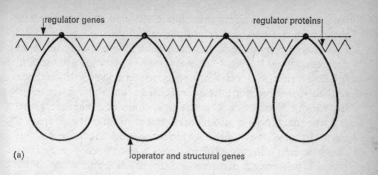

regulator genes   regulator proteins

(a)

operator and structural genes

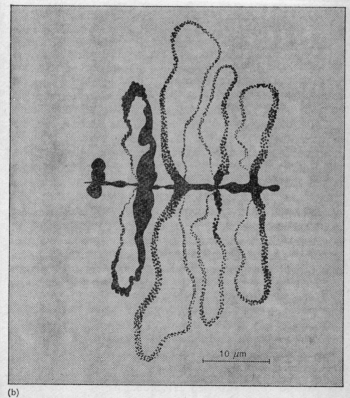

10 μm

(b)

parts of the body in harmony. The possibility of studying many of these processes has only recently come above the scientific horizon but, now that a start has been made, there should be rapid progress.

There are also special problems to be solved, which have probably now become accessible, such as how differentiation of the various types of cells occurs in a complex organism. There must be many points in the development of an embryo, from a single fertilized egg cell, at which one cell gives rise to two daughter cells differing in character. These differences are maintained in subsequent divisions and ultimately lead to the formation of groups of cells of different characters such as occur in special organs.

As both daughter cells have identical sets of genes, the differences must be due to an uneven distribution of the cytoplasm – non-nuclear material – between the two. For example, one daughter cell may contain substances which activate one set of genes while the other may possess different substances which react differently. It is known that in the initial divisions of fertilized egg cells, the substances present in the cytoplasm become unevenly distributed among the different cells and, as a result, initial differences of gene activity in the different cells may be established. It is far from clear how these differences are perpetuated. It could be that, having been initially activated, certain genes themselves produce substances which ensure that their activity is maintained in subsequent cell generations. Such genes, once activated, will be self-activating so that one type of cell will perpetuate itself – liver cells once formed will continue to produce liver cells.

These problems are profoundly difficult, but until they are solved it cannot be claimed that we know much about the principles on which complex organisms are constructed. This knowledge of control is also a prerequisite to an understanding of those cases in which the controlling mechanisms have broken down, as in malignant tumours.

Although an enormous amount of research has been carried out on cancer and the circumstances which give rise to it, the mechanisms by which a normal cell is transformed into a tumour cell have not been discovered. It is probable, though not certain,

that in many cases the abnormality in the cancerous cells is the result of an actual change in the genes. This would cause the modification of the original cell and its progeny in such a way that they escape from the normal controls. A large number of agents – X-rays, intense radiations and numerous chemical compounds – have been found experimentally to produce tumours in animals. Many of these substances have been shown to react with the units of DNA which are responsible for the genetic code and will therefore give rise to some abnormality in it. In some instances they may even cause a deletion of part of the code; the tumour could then be regarded as originating in a cell in which some of the genes which control cell growth and multiplication have been lost. Until these controlling processes have been brought into the open it is unlikely that there can be any real understanding of the circumstances under which this loss of control occurs. In this sense it can be said that the real scientific study of the nature of cancer can only now begin.

This situation is a good example of the rule, which has occurred so often in the history of science, that progress cannot be made in elucidating a complex phenomenon until the necessary basic facts which must come into the solution have been recognized. The origin of the energy given out by the sun and stars remained totally obscure until the discovery of radioactivity led to the recognition of nuclear reactions and nuclear energy. Nothing was known about the relations between the chemical elements until means of probing the atomic nucleus were available. Little could be established about the mechanisms of heredity until the substances concerned, nucleic acids and proteins, were characterized.

In the same way, the complex and beautiful control mechanisms which permit hundreds of types of different cells to cooperate in producing a single organism have only recently become accessible for study because some of the main mechanisms which are common to all cells have only just been elucidated. It might be added that only in the past two decades have the instrumental means of such studies been available – the use of radioactive compounds to trace the transformation of substances and metabolic pathways within cells and organisms and the greatly improved

methods of examining and isolating cell structures now available in the electron microscope and the ultracentrifuge. Without these progress would have been much slower and could hardly have reached a stage at which the problems formulated in this article could have been usefully discussed at all.

7 G. Chedd

Genetic Control: the Debate Burgeons

G. Chedd, 'Genetic control: the debate burgeons', *New Scientist*,
7 March 1968, pp. 512–14.

Nothing is so intriguing to an interested but impartial observer of
the scientific scene than to watch the slow unfolding of two rival
theories. With all the fascination of a game of chess, and with
stakes as high as the understanding of one of the most fundamental problems facing biologists today, the debate as to the
method by which cells control themselves continues to escalate in
strict accordance to the gospel of Herman Kahn. One new piece
of experimental evidence on one side is immediately matched by
a new result from the rival group. Just as a checkmate seems
inevitable, a fresh experimental pawn sidles up from nowhere and
the game opens up again. Both players' kings, the cores of their
respective theories, are now guarded by a formidable array of
such experimental pieces.

The two groups are trying to crack open the problem of cell
differentiation. Every cell of an organism contains an identical set
of genes; the genes control what each cell is to be (a component
of a sycamore tree or a duck-billed platypus or even your Aunt
Edith) by directing the synthesis of the cell's enzymes; the enzymes in turn carry out the orders of the genes by virtue of their
ubiquitous influence on every bit of the chemistry that goes on
in the cell. The chain of command that makes up this control
pathway is now understood in considerable detail. The genetic
code, the master-plan, carried within the structure of the DNA
molecules that constitute the genes in the cell's nucleus, is first
copied, in short sections, on to molecules of RNA, which are
chemically similar to DNA. The instructions, still codified, but
now carried by RNA, are then transported out of the nucleus
and into the surrounding cytoplasm by these messenger molecules. Here, on small particles known as ribosomes, the code is

translated by a highly complex process, only the bare outline of which is yet understood, into the cell's enzymes.

So far, so good. But this elegant scheme makes no provision for the obvious fact that any one organism contains many different kinds of cell. Nerve cells, for instance, look and behave very differently from the cells that make up, say, the lining of the stomach, yet, they all contain the same genes. The difficulty can be got around by proposing that in different types of cell different sections of the genes' DNA are being read and translated into enzymes. Indeed, in any one cell it appears that some 99 per cent of the genetic information is not being expressed. In other words, only a small minority of the genes are actually working; all the rest are switched off.

Another aspect of the same problem is that cells are able to react to their environment. Genes that are switched off can be switched on, and vice versa. Under normal conditions, their ability to do this is fairly limited. A brain cell can't suddenly become a skin cell, for instance, by switching off all the genes that make it brain-like and switching on those that would make it skin-like. There is one only too well known example of the consequences of a catastrophic breakdown of the normal strict genetic regulation – cancer.

The fact that the genes of a cell will react to their environment by a shift in the pattern of messenger-RNA synthesis can be, and has been, demonstrated in several ways. Two will be discussed in some detail here; both are very different in the experimental systems used, but both imply that some substance from outside the immediate environment of the genes can move into their vicinity and switch them on or off. The reason for making these particular choices from among the many examples will become transparently obvious later.

The bacterium *Escherichia coli* only makes certain enzymes when they are wanted. If grown in the presence of the sugar lactose, the bacteria produce enzymes that will break the sugar down and allow the bacteria to use lactose as a source of carbon and energy. Grown in the absence of lactose, the bacteria do not synthesize the lactose-splitting enzymes. As a result of a series of brilliant experiments, Jacob and Monod of the Pasteur Institute in Paris constructed a hypothesis to explain the admirable

economy of *E. coli*, a hypothesis that involved postulating the existence of a molecule known as the *repressor*, which could 'turn off' the genes responsible for the lactose-splitting enzymes.

Jacob and Monod's hypothesis – which now constitutes one of the jealously guarded kings on the chess board – was published in 1961, and had an immense impact on molecular biology; so great, indeed, that they were awarded a share of the Nobel Prize for Physiology or Medicine in 1965. Their ideas assumed that the hypothetical repressor molecule is sent out by a so-called *regulator* gene; the repressor can then bind with a nearby gene known as the *operator* gene, which is next door to the genes controlling the lactose-splitting enzymes. When the repressor is bound to the operator gene, the adjacent working genes are switched off; when repressor is not bound to operator, the genes are switched on (in other words, they can be copied by messenger RNA, which in turn is translated into the relevant enzymes). What prevents the repressor from binding with the operator, so allowing lactose-splitting enzymes to be made? Why, lactose itself, which is assumed to combine with the repressor molecule and prevent its interacting with the operator. To sum up, lactose presses the button that starts its own destruction by combining with the repressor substance put out from the regulator gene. The repressor then cannot, as it would in the absence of lactose, combine with the operator gene. The enzymes can therefore be synthesized.

Elegant though this hypothesis was, it was only a hypothesis. The experiments that led to its formulation were entirely genetic; the interaction of repressor and operator, the interaction of repressor with the lactose (called the inducer) indeed the very existence of the repressor itself – were all only postulates. Jacob and Monod's king was on the board, but only very lightly guarded, and it made a very tempting target for the attackers.

A secondary theory postulating the external control of genes arose from the well-known experiments of Henry Harris, Professor of Pathology at Oxford. In 1965, he succeeded in making hybrid cells, cells that contained the nuclei of two entirely different species: man–mouse cells, for instance. Among the several important consequences of the experiments, in which both species of nuclei in a single cell keep working, was that the cytoplasm recognized signals from a 'foreign' nucleus. More important still, in

our context, he showed that the foreign nucleus recognized signals from the cytoplasm.

Certain cells in vertebrates reach such a high degree of specialization that all their genes, with the exception of just one or two, are switched off. The red blood cells in birds are a case in point; they make no DNA, and only a very, very small amount of RNA. Harris introduced bird erythrocyte nuclei into human cells that were actively synthesizing DNA and RNA, and found that within twenty-four hours RNA synthesis in the bird nucleus was appreciable. In other words, signals from the human cytoplasm had entered the bird nuclei and switched on their genes: bird nuclei had recognized human cytoplasmic signals.

Harris argued that his experiments were incompatible with Jacob and Monod's repressor hypothesis. He proposed, instead, the existence of an *activator*. The suppressor hypothesis hinges on a specific, negative mechanism; Harris's activator is non-specific and acts in a positive sense. Moreover, observations of the chromosomal material of the bird nuclei suggested to him a mechanism by which his activator might work. Before introduction into the human cell, while its genes are all switched off, the chromosomal material (chromatin) is dense and the whole nucleus is very small. However, after introduction the nucleus enlarges several times and the highly condensed chromatin opens up. Harris suggested, then, that the activator causes the nucleus to enlarge and allows the genes to become much more accessible to the enzymes that allow RNA to be copied.

The second king had been placed on the board. Like the rival theory, however, it was still vulnerable. Although explaining how large numbers of genes are switched on, it could not explain how *specific* genes are activated. Moreover, it was still in the same embarrassing position as the Jacob–Monod hypothesis, because, although necessary for the scheme, the existence of the activator was still only a postulate; there was no physical proof for its existence. And there, with the two kings glaring at each other across the board, the situation remained.

The first side to move into action, and begin the experimental escalation, was the Jacob and Monod school. The bulk of experimental support has come not from Paris but from Harvard, where two teams, one headed by Gilbert and Müller-Hill, the

other by Ptashne, succeeded late in 1966 in finding actual repressor molecules. Ptashne isolated a repressor from a virus, while Gilbert and Müller-Hill found the 'original' repressor (that is, the one which operates on the genes that direct the lactose-splitting enzymes in *E. coli*). Moreover, the method used by Gilbert and Müller-Hill also showed that the repressor binds to the lactose inducer. Two vital experimental pawns placed in one move.

Next into action was Harris himself, with experiments performed at about the same time but published later. In these experiments he showed that the amount of RNA made in the bird red blood cell nucleus – the degree to which it is switched on – is related to the degree of nuclear enlargement. Moreover, he claims to have found in these experiments 'nuclear bodies' suggesting that the opening up of the condensed chromatin is not a random process. 'It is thus possible,' he writes, 'to envisage a system of genetic regulation in which changes in nuclear volume are associated with the opening up or closing down of specific areas of chromatin in ordered sequence.' He claims that the cytoplasmic signals, 'could thus regulate not only the amount of RNA made in the nucleus but also the areas of DNA on which it is made.' This somewhat shaky experimental pawn therefore gets around one problem of his theory; however, the cytoplasmic signals – the activators – remained molecules of convenience.

The next two moves in the game were also made at about the same time, and both have been published within the last few weeks. First Ptashne and then Gilbert and Müller-Hill succeeded in showing that their respective repressors do indeed bind with a section of DNA. The latter workers, moreover, estimate that each *E. coli* contains ten to twenty repressor molecules, and that each molecule must recognize a stretch of DNA about 35 Å long. From their earlier experiments they estimated that the molecular weight of the repressor molecule is about 150 000, which is quite large enough to cover this length of DNA.

Meanwhile, support for Harris was coming from a new quarter, the University of Washington, where Thompson and McCarthy confirm that cytoplasmic signals can switch on the genes in dormant nuclei. They have shown this by incubating isolated hen red cell nuclei with the cytoplasm of mouse liver cells.

In unpublished work, they have also shown that human cytoplasm works just as well in switching on dormant bird nuclei, confirming too that the cytoplasmic signals are not specific.

The most important part of their paper, however, is the last paragraph, where they claim to have partially purified the factor that stimulates RNA synthesis. No details are given, as the experiments are still in progress and will be reported elsewhere. However, a new and very powerful defensive pawn – maybe even a rook – is hovering just above the board in front of Harris's king.

The latest move in the game, reported in *Nature*, once again comes in support of the repressor hypothesis, and once again comes from Harvard, though from a new team headed by Karin Ippen and including John Scaife, on leave from the MRC Microbial Genetics Research Unit in London. Their paper gives an important insight into how the binding of the repressor to the operator gene switches off the adjacent 'working' genes. For RNA to be copied on DNA, an enzyme called RNA polymerase is needed which, it is envisaged, works like a zip, attaching to the DNA stand and moving along it, making RNA as it goes. The Harvard team have identified the region of the gene where the zip becomes attached when making the messenger RNA for the lactose-splitting enzymes. The operator gene turns out to lie between the point of attachment of the polymerase enzyme and the working genes that the polymerase copies. This discovery immediately suggests a simple mechanism by which the repressor prevents the copying of these genes: when attached to the operator, the repressor, like a piece of knotted cotton, simply blocks the progress of the zip.

And there the game rests, apparently at stalemate. It can of course be argued that the very different experimental systems used by the two sides – one concerned only with bacteria and viruses, the other with vertebrate cells – invalidate the game, each mechanism applying only to one system.

However, Jacob and Monod's supporters would be very reluctant to concede that the repressor hypothesis is restricted to bacteria and viruses, while Harris himself implicitly acknowledges that Jacob and Monod's scheme might be applied to higher organisms by losing no opportunity to attack it in his papers, despite these being only concerned with the control of genetic

activity in multicellular animals. Both sides' serried ranks of pawns now seem so impenetrable that it is doubtful whether either king will be taken. The usual outcome of such a scientific stalemate is that a new theory will emerge, incorporating both the present theories.

8 B. Afzelius

Lysosomes: Bags of Digestive Enzymes

B. Afzelius, *Anatomy of the cell*, ch. 8, 'Lysosomes; bags of digestive enzymes', University of Chicago Press, 1966, pp. 60–64.

The term lysosome has been constructed from Greek word roots and means 'digestive body'.

The concept behind the term, not to mention the reality behind the concept, is much debated. De Duve, who coined the term, gave it a concrete biochemical meaning: a cytoplasmic particle containing several digestive enzymes which are prevented from acting on the cell machinery by some kind of barrier, presumably a limiting membrane.

A lysosome can thus be regarded as an organelle which performs much the same functions in the cell as do the digestive organs in the body of a higher organism. Since a cell is simpler than an entire organism, the analogy cannot be expected to be very close. The lysosome has many abilities which a stomach lacks and vice versa.

De Duve used the distribution of many enzymes among various cytoplasmic fractions obtained by density centrifugation as an experimental basis for the lysosome concept. [. . .] He found catalase, uricase, and D-amino acid oxidase to be localized at a specific level which was later found to contain the particles which had been called microbodies. In another fraction there were several enzymes with common properties: they catalysed reactions in which compounds were split with simultaneous addition of water and had maximal enzymatic activity at an acid pH value. Such enzymes are called acid hydrolases, and De Duve conceived the idea that the cytoplasmic fraction contains bags of acid hydrolases. These bags he called lysosomes.

The concept was based on some assumptions. One assumption was that a single lysosome contains not one, but several, enzyme species. This was not supported by any direct observation but

seemed likely from various lines of reasoning. The particular attraction of this assumption is that the lysosome would then be a complete digestive bag. Whatever fragment of a cell should come into the digestive bags, the chances are that it would be degraded into its smaller components. A complete digestion could not be performed if there were as many lysosomal species as there are lysosomal enzymes.

Another assumption was that the hydrolases are all in an active state in the bag and are kept from digesting the cell only by their limiting membrane and perhaps an unsuitable pH value. But why don't they digest each other? The enzymes are proteins, and some enzymes in the lysosome are protein-digesting (proteolytic) enzymes and can thus be dangerous to the neighbouring enzymes in the lysosomes. The problem of how the enzymes are protected from each other in a lysosome is not an easy one. Perhaps the pH value in a resting lysosome is far from the optimal value for a proteolytic enzyme; perhaps the enzymes of the lysosomes are unusually resistant to the action of proteolytic enzymes. There is also the possibility that the lysosomes do degrade themselves slowly and have to be formed anew continuously. The latter possibility is not likely, however, since the average lifetime of a lysosome from the rat liver has been found to be fifteen to thirty days, which is about twice the calculated value for a mitochondrion from the same tissue.

Today, more than ten years after the creation of the lysosome concept, there is nothing in the original concept that has had to be changed. Not only the lifetime of the lysosome, but also the lifetime of the lysosome concept is quite respectable.

Yet there have been attempts to redefine the lysosome. The electron microscopists in particular have been tempted to apply the term lysosome to particles which probably, but not necessarily, are identical with lysosomes as identified by biochemical criteria. The electron microscopist has the advantage over the biochemist in that he can examine the particles one by one, but he is far less well equipped with methods which enable him to detect the enzymatic properties of the particles. Some lysosomal enzymes can, however, be detected with methods which can be used in electron microscopy, the easiest among them being for the enzyme acid phosphatase. There has therefore been a tendency to create a new

definition of lysosomes for electron microscopy: a lysosome is a particle, with a single limiting membrane, in which acid phosphatase is demonstrable.

Much discussion is at present devoted to the question of how far the electron-microscope lysosomes are identical with the biochemically defined lysosomes. They would be the same if every lysosome contains acid phosphatase which can be demonstrated by the electron microscopist and if every particle containing acid phosphatase is also equipped with the other dozen lysosomal enzymes. According to present evidence, the first condition is always found to be true, but the second condition is not. The term lysosome is likely to cover more structures when used in connection with electron-microscope studies than when used with biochemical work.

It is important to elaborate methods for localizing the other lysosomal enzymes in electron-microscope sections. Much ingenuity is now being devoted to these problems, and it is likely that in the near future it will be known whether a single particle contains the full set of lysosomal enzymes, whether there are lysosomes of different kinds, and whether lysosomes are a general cell component.

What then does a lysosome look like? [. . .] There is no standard appearance for a lysosome. Lysosomes are known to differ in size and shape from one cell type to another and within the same cell type under various functional conditions or at various developmental stages during the life of the lysosome.

The lysosomes start functioning when the cell takes up substances either by phagocytosis or by pinocytosis. The lysosomes then fuse with the packages of ingested material, and the digestive enzymes of the lysosomes attack the digestible material. The uptake of the material by pinocytosis or phagocytosis may last from a few minutes to a few hours; and the digestion within the fused bags may require a few hours to a few days, depending on the nature of the ingested material. It seems that the cell shows no discrimination in this process: the same complement of enzymes are released into the bags of ingested material regardless of whether the bags contain a more protein-rich or a more carbohydrate-rich material, or whether the cell was fooled into ingesting a completely insoluble fluid or particle. In this respect the cell

machinery is relatively automatic with a low degree of flexibility and inferior to the digestive organ of a higher animal.

Digestion within a lysosome, or a lysosome fused with an ingested vacuole, normally gives rise to amino acids, sugars and simple organic compounds which might be useful building blocks in the synthetic machinery of the cell. It is therefore somewhat surprising that some cell types do not use the building blocks formed in the process themselves but expel them from the cell. Other cell types, however, such as the protozoans, rely entirely on this mechanism for their metabolic needs.

Insoluble remnants usually stay within the lysosome, which in this way becomes packed with different inclusions. Lipids often become inclusions because they are not digestible by the lysosome which may lack lipid-degrading enzymes. Fatty substances accumulating in the lysosomes are often colored and are sometimes called wear-and-tear pigment. The lysosomes containing them are called lipofuscin granules. They characterize, above all, cells which have lived and functioned during a long period, for instance those in the muscle or other tissues of aged people or animals. In the protozoans, which are given the gift of eternal life, the cells are capable of getting rid of the bags of insoluble matter by a defecation-like process. The same is also true of the cells of the kidney and some other organs.

The processes described above have the uptake of material from the outside as a starting point. This is not necessarily the only impetus for the formation of a functioning lysosome. There are a few cases described in which the lysosomes start to digest the cell in which they reside. In one particular case, only a limited region of the cell is encapsulated and a lysosome is formed around this region or out of this region in an hitherto obscure manner. This may, for instance, happen when an animals is starved. The mechanism gives the impression of serving the purpose of sacrificing one part of the cell in order to give nutritive substances to the rest. There are also cases in which the lysosomes open their bags of digestive enzymes and kill and digest the cell. This may happen when an embryonic organ, for example, the oviduct in a male chick embryo, is to be resorbed.

One wonders whether the lysosome bags might open accidentally and thereby damage the cell. This may occur upon exposure

to several poisonous substances. There may also be a difference between the action of substances which induce the lysosomes to burst and thereby lose their enzymes and substances which only slightly change the membrane of the lysosomes so that they remain relatively intact but allow their content of digestive enzymes to leak out slowly. The membrane of the lysosome can also become more stable when it is treated with a substance, for example, with the hormone cortisone, which includes among its actions a direct one upon this cell organelle. Finally, there may be inborn defects of the lysosomes – which has been suggested for at least three diseases.

One of the main revolutions in biological thinking was the abolition of the old humoral pathology and its replacement by cellular pathology. Today it is possible to carry the revolution one or two steps further and to seek the cause of disease in disorder at an organelle, or even a molecular, level.

9 A. C. T. North

The Structure of Lysozyme

A. C. T. North, 'The structure of lysozyme', *Science Journal*, vol. 2, 1966, no. 11, pp. 55–63.

Lysozyme was discovered by Fleming in 1922, six years before he found penicillin. Suffering from a heavy cold, he tested the effect of his nasal mucus on a bacterial suspension and found that the opaque suspension rapidly became clear. Further investigation revealed that the mucus contained an enzyme which could break down – or lyse – the cell walls of the bacteria and Fleming called the enzyme lysozyme. He hoped that lysozyme would be a useful weapon against all sorts of bacterial infection, but unfortunately it is only effective against certain types of bacteria and these are generally fairly harmless. However, their very harmlessness may be due to the presence of lysozyme, which is to be found in body fluids, including tears and sweat and egg white.

An enzyme is a protein molecule whose function is to catalyse a chemical reaction in a living system. Its substrate is the principal reactant which is modified as a result of the reaction. The structural relationship between an enzyme and its substrate involves a very close and specific matching of molecular surfaces and attractive groups. It is indeed rather like a complicated lock and its key, since it is necessary for the enzyme to be able to distinguish between chemical groups that are generally quite similar.

In order to understand such a mechanism, one must therefore know both the actual linear sequence of the amino acids along the protein chain and how the chain itself is folded up in space.

The amino acid sequence of protein molecules can be worked out by relatively straightforward chemical analysis and in lysozyme was demonstrated by Jolles and Canfield in the early 1960s. However, the use of physical methods – notably X-ray diffraction techniques – is required to enable the spatial

arrangements of all the atoms in the molecule to be determined. This was first achieved in 1960 for the respiratory protein myo-globin – which acts by combining reversibly with molecular oxygen and so storing it in the muscle tissues. During the past five years Phillips, Blake and myself with our colleagues at the Royal Institution have successfully elucidated the three-dimensional structure of lysozyme. This molecule, which comprises a single chain of 129 amino acids, is thus the first enzyme and the second protein to have been completely analysed.

In order to see the structure of a molecule in atomic detail it would be most convenient if one could examine a single molecule under a high-powered microscope. This is technologically impossible, but fortunately many proteins may be crystallized. Such crystals are composed of regularly repeated and identical 'unit cells'. Each cell may contain either a single molecule or several molecules arranged symmetrically within it. The structure of a crystal, and hence its unit cells, can be investigated by the technique of X-ray diffraction.

Essentially, this technique depends on the fact that the clouds of electrons surrounding atoms will scatter X-rays, so giving a measure of the electron-density distribution within the molecule; concentrations of electrons pin-point the position of individual

Figure 1 The X-ray diffraction technique used to determine the three-dimensional structure of lysozyme is illustrated here. The crystal is used as the target of a beam of X-rays which are reflected by the electrons of atoms lying in the crystal planes. The reflected X-rays are recorded photographically and form a symmetrical pattern of bright spots – a diffraction pattern. The amplitude and phase of a reflection depend upon the way that the electrons, and hence the atoms, are distributed between the reflecting planes. The amplitudes can be measured directly but the phases are determined using the isomorphous replacement method (see Figure 2). Three-dimensional electron contour maps can then be constructed. If data derived only from reflections near the centre of the diffraction pattern are used, a low-resolution map is obtained which does not contain fine detail. From this a three-dimensional model can be constructed which shows the general conformation of the molecule. However, to make an accurate, high-resolution model, reflections from the whole of the diffraction pattern must be used to give a much more detailed electron-density contour map. The high-resolution model of the lysozyme molecule thus obtained is shown (bottom, right)

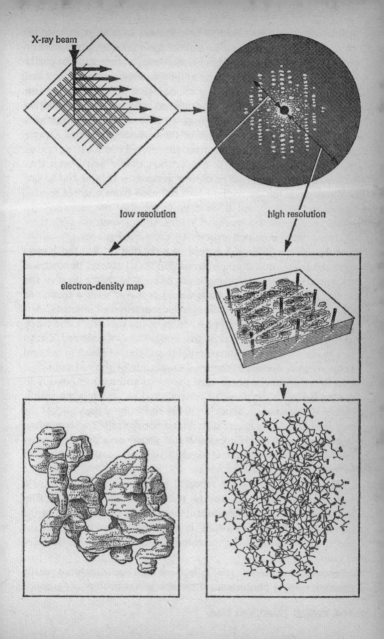

X-ray beam

low resolution

high resolution

electron-density map

atoms within the structure. When a beam of X-rays of a single wavelength is directed at a crystal the successive similar planes of atoms within it behave as though they were a succession of partly reflecting mirrors – each plane adding its own quota to the total amount reflected. The fraction of the beam reflected by the top plane will be reinforced by the fractions reflected by the underlying planes if the additional distance that those other fractions must travel is an exact multiple of the wavelength of the X-rays. In such a case, the reflections from the second plane will be one or more complete wavelengths behind that of the first plane; that from the third plane two or more wavelengths behind and so on.

The total intensity of the beam reflected from a set of crystal planes in this way will depend upon the way the electrons are distributed between successive planes. If all the atoms with their electrons were confined exactly to the set of planes the corresponding X-ray reflection would be a maximum. But the X-rays reflected from additional electrons half way between the original set would be half a wavelength out of step, so that a crest of the additional reflected wave train would coincide with a trough in the original component, leading to a diminution of intensity. Additional electrons some other fraction of the way from one plane to the next will again give rise to a component of reflected X-rays that is out of step with the original train; this will lead in general not only to a change in the total amplitude of the reflection but also to a shift in the position of the crests and troughs, that is to say, a change in the phase of the reflected wave. Thus the amplitude and phase of a particular X-ray reflection depend upon the way that the electrons are distributed between the corresponding reflecting planes. The amplitude and phase of a different X-ray reflection depends upon the electron-density distribution in another direction in the crystal.

The set of X-ray beams reflected by the crystal is known as a diffraction pattern and may be recorded on photographic film (see Figure 1). To extract information about the structure of the crystal from these X-ray data, it is necessary to know both the amplitudes and phases of the complete set of reflections from the crystal.

The amplitudes of the reflections can be measured easily enough, using a photometer if the image has been photographed

or a Geiger counter if it is being studied directly. There is, however, no direct method of measuring the phase.

Various ways of overcoming this difficulty can be used for elucidating the structures of relatively simple molecules, but the problem remained intractable for large molecules, such as proteins, until Perutz developed a technique which had previously been used mainly in the determination of the structures of relatively simple crystals. Known as the isomorphous replacement method, this involves attaching an additional atom – such as mercury or uranium – to each molecule in the crystal. The atoms chosen are ones which reflect X-rays strongly.

An isomorphous substitution, therefore, will lead to a change in the strength of the X-ray reflections and by comparing the two sets of intensity data – one from the untreated or native crystal and one from the substituted crystal – it is possible to deduce the location of the substituted atoms. When their positions are known, their contributions to any reflection can be calculated both in amplitude and phase. It is now possible to deduce the phase of the reflections from the other atoms because for any one reflection we know the amplitudes but not the phases of the reflections from the native crystal; the amplitudes but not the phases of the reflections from the substituted crystal; and both the amplitudes and phases of the reflections from the substituted atom.

Using this information it is possible to arrive at two alternative solutions to the phase problem, of which only one is correct. To overcome this difficulty a second isomorphous compound must be prepared with the substituted atom in a different place. This will give a second pair of solutions for each phase, of which one will be the same as one of the original values. In practice, it is advisable to have three or even more substituted derivatives because each one will inevitably make only a weak and unhelpful contribution to some of the reflections.

This technique of substituting is not easy. It is essential that the modified molecules crystallize with exactly the same packing arrangement as the normal unmodified molecules so that the only difference between the two crystals is that the reflecting power is significantly greater in the modified crystals at one place in each molecule. This can be achieved in two ways. First, it may be possible to carry out a chemical reaction in solution between

the protein molecule and a molecule containing the heavy atom, so that a compound is formed. The difficulty with this technique is that the additional group may either modify the packing arrangement or even prevent crystallization altogether. Secondly, it may be possible to diffuse small molecules containing the heavy atom into protein crystals that have already been grown. This is less likely to result in packing changes but the binding is often not sufficiently specific, or else there may be a number of binding sites. However, it is possible to cope with a small number of sites and this method has proved the more successful, though up to now completely satisfactory substitutions have been achieved only in studies of some half-dozen proteins.

So far, I have not discussed how many reflections will have to be measured to give the required image. This depends upon how high a resolution is required. Using only reflections near the centre of the diffraction pattern will give a picture that is broadly correct but which will not contain fine details. To see the fine details, higher-angle reflections from the outer part of the pattern must also be used.

The low-resolution image will reveal only the broadest features of the molecule but even this may be sufficient for some purposes. It has been the first goal in each of the structure determinations carried out so far. For a small protein such as lysozyme, a low-resolution image may require the measurement of some 800 reflections but for a resolution of 2 Å, sufficient to reveal chemical groupings such as amino acid side chains as distinct features, some 20 000 reflections must be measured. Every reflection must be measured for the native protein and each heavy-atom derivative. Unfortunately, crystals are liable to deteriorate as a result of radiation damage and in order to minimize this effect it is necessary to restrict the number of measurements that can be made from any one crystal and to use a number of crystals. A certain amount of overlap must then be allowed in order to relate data from different crystals and, allowing for unsuccessful runs, something like 250 000 measurements will probably be required altogether.

The need to make such a large number of measurements has brought about the development of automatic apparatus which

scans the diffraction pattern, moving crystal and X-ray detector through appropriate angles, and records the intensity on computer tape or feeds the data directly into an on-line computer. All the subsequent stages of handling data are carried out by computer.

The output of the computer comprises values of the electron density throughout each unit cell of the crystal. These densities are usually plotted in the form of a contour map for a particular cross-section of the unit cell. A set of such maps drawn on transparent sheets can then be stacked together. This three-dimensional map has then to be interpreted in terms of the known chemical composition of the molecule. Since a protein molecule is known to have a main chain with side chains branching from it at regular intervals, the electron-density map ought to show a continuous ribbon of density with branches; the maps of myoglobin and lysozyme have shown just that.

Given an electron-density map, the next step is to construct a model of the structure. This requires the use of results that have emerged from studies of small molecules for which atomic resolution can readily be achieved. For example, bond lengths between atoms that are covalently linked tend to be constant for any particular pair of atoms and when an atom is linked to two or more other atoms, the inter-bond angle tends to be fairly constant.

The structures of all the individual amino acids – and of many combinations of two or three amino acids – have been worked out. While there is some flexibility in some of the side chains, the permissible atomic arrangements are rather limited. It is therefore possible to make a kit of parts with rigid pieces representing those features that are well defined and constant, and with joints allowing flexibility where appropriate. In constructing a model from an electron-density map of a protein, it is of great assistance if its amino acid sequence is known for, whereas some of the side chains can be readily identified by their characteristic 'electron' shape, some have rather similar shapes and are extremely difficult to identify unequivocally.

The model-building procedure is first to identify a particularly well-defined feature in the map and measure its position; secondly to place the corresponding model part in the right place in a supporting framework. The positions of connected atoms in the

model can then be measured and the map examined to see whether these atoms fall within the electron-density peaks. If not, the model has to be adjusted until they do. Some other rules are important – for example, there are well-defined minimum values for the distance between atoms that are not bonded to each other – and the model has to be checked to see that these have not been violated. The model-building process is rather time consuming and it is likely that, in future, computers will be used at least to carry out the adjustment of model components to fit the observed density. A difficulty in allowing the computer to do the whole job is that departures from the rules do occur from time to time and it is important to be able to allow some flexibility where the circumstances warrant it. It would be rather difficult to draw up a definite list of such circumstances in advance.

Using the techniques I have just described, we were able to construct the first complete model of the lysozyme molecule early in 1965. We saw at once that it bore only modest similarity to that of myoglobin.

The angular relationship between two successive units of a protein main chain is restricted to fall, within fairly narrow limits, into one of a few classes, in order that the non-bonded atoms in the two units do not come too close together. If several successive units bear the same angular relationships to each other, the chain becomes coiled. One such arrangement, known as the α-helix, was found to occur in a number of synthetic amino acid polymers, in fibrous proteins such as wool and in myoglobin, which is about 75 per cent α-helix. The myoglobin chain contains eight lengths of α-helix linked by comparatively short non-helical sections of chain at the corners; these allow the complex molecule to be folded up into a compact volume.

However, much of the main-chain conformation of lysozyme appears to be a rather irregular succession of the different possible conformations and the α-helix does not predominate. Indeed, it accounts for only 25 per cent or so of the chain, mainly in three lengths of about ten amino acids each. The only other conformation that is developed to any extent is one in which the chain is extended almost as fully as possible. This arrangement has been found in fibrous proteins, such as natural silk, in which there are

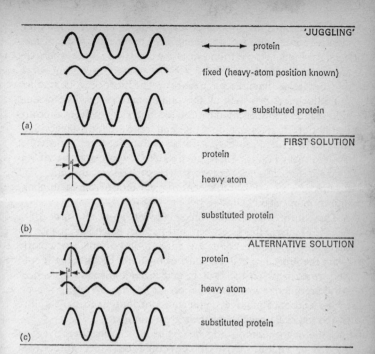

Figure 2 An isomorphous replacement technique is used to determine
the phases of the X-rays reflected from the crystal. Developed by
Perutz, the technique involves attaching an additional heavy atom —
such as mercury or uranium — to each molecule in the crystal. The
amplitude and phases of a reflection from the substituted protein is the
sum of the separate contributions from the protein and the heavy atom,
each with its characteristic amplitude and phase. Since the heavy atom
reflects X-rays strongly, it is possible to fix its position accurately and
hence deduce the amplitude and phase of its contribution. By shifting
or 'juggling' the phases of the protein contribution and the total relative
to the heavy atom (a) it is possible to find the phases for which the
contributions add correctly together. Unfortunately two results are
obtained depending on whether the crests of the heavy-atom
contribution lie to the right or left of the protein contribution (b and c).
To overcome this difficulty a second substituted crystal must be prepared
with the heavy atom in a different place. This will give a second pair of
solutions, of which one will be the same as one of the original values

many chains arranged parallel to each other in sheets, alternate chains running in opposite directions. This anti-parallel arrangement is, in fact, to be found in lysozyme at one place where the chain doubles back on itself.

What holds the molecule together? The most obvious stabilizing influence is provided by the amino acid cystine. Two cystine side chains in different parts of the main chain can come together to form a disulphide bridge and there are four such bridges in lysozyme (see Figure 2). Another stabilizing force is provided by hydrogen bonds. These are relatively weak bonds that are formed, for instance, between NH or OH groups and CO groups; such groups are present in each unit of the main chain and in many side chains.

They form longitudinal links between adjacent turns of a helix and lateral links between anti-parallel extended chains. Individually, hydrogen bonds are not very important, particularly when the substance is in an aqueous medium, because they could be formed between the protein and water molecules instead of between different groups of the protein. Cumulatively, however, hydrogen bonds make a significant contribution to the stability of the molecule because of their great number.

Some of the amino acid side chains are charged negatively or positively, and attractive forces would be expected between groups having opposite charges. However, only one such pair of groups are juxtaposed in lysozyme. All these polar and hydrogen-bond forming groups are distributed over the surface of the molecule. There remains a further class of side chains which are hydrophobic or greaselike. They are repelled by water and therefore tend to point towards the interior of the molecule. A number of

---

Figure 3 The catalytic activity of an enzyme appears to be situated in a deep cleft down one side of the molecule into which the cell wall or substrate fits. So far, experiments done with a cell-wall 'analogue' – tri-*N*-acetyl glucosamine (tri-NAG) which contains three sugar rings – show that it fits into the positions marked A, B and C. A longer polysaccharide than tri-NAG, such as actual cell-wall substrate, would be expected to extend further along the cleft as shown here, but it would be surprising if the type of experiment carried out so far showed a substrate molecule bridging across the catalytic site, because any molecule that did so ought to be cleaved. It seems most likely, in fact, that the actual catalytic site lies between sites D and E

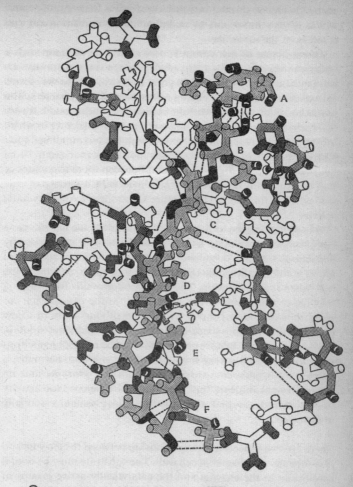

Symbol	Legend
⊖ carbon–hydrogen group of enzyme	main chain of enzyme
carbon–hydrogen group of substrate	carbon atoms of substrate
oxygen atoms	side chains of enzyme
nitrogen atoms	bonds

such side chains coming together can form a hydrophobic core, which is very important in maintaining the conformation and stability of the molecule.

The structure of myoglobin is very clearly based upon such a hydrophobic core, with hydrophilic groups on the surface together with a few hydrophobic ones. Lysozyme, however, seems to consist of two regions having rather dissimilar characters. The first is compact, with a hydrophobic core, but the second has very few hydrophobic side chains and a rather open structure in which most of the side chains are in contact with the surrounding liquid. There are even at least two water molecules that seem to be hydrogen bonded within the protein. Between these two wings of the structure is a deep cleft that is lined with a variety of side chains. It is this cleft that appears to be the site of enzymatic activity.

It is of importance to know whether the molecule has the same conformation in solution as it does in the crystal; otherwise any deductions that might be made from the structure in the crystal might not be a valid guide to the functioning of the enzyme in the cell, when it is in solution. There are good grounds for believing that the conformation is the same. First, in the crystal each enzyme molecule is in contact with its neighbours only in a few places. The remaining space is filled in with the aqueous solution from which the crystals are grown and thus the molecular environment in the crystal is very similar to that in solution. Secondly, a number of experiments carried out both on lysozyme and on other enzymes indicate that they retain their enzymatic activity in the crystal, so that the structure must remain essentially undistorted.

There are two aspects to the relationship between the enzyme and its substrate – the bacterial cell wall. The substrate must be bound specifically to the enzyme and the catalytically active groups of the enzyme must react with the substrate.

An essential constituent of bacterial cell walls is a polysaccharide chain consisting of alternating units of N-acetyl glucosamine (NAG) and N-acetyl muramic acid (NAM). Both these units are formed from the sugar glucose by the addition of chemical groups. Short tails consisting of four amino acids are attached

to the NAM units and link across to similar tails from a neighbouring chain, so that a complicated meshwork of cross-linked chains results.

The action of lysozyme is to cleave, or hydrolyse, the bond between an NAM unit and the oxygen attached to the next NAG so that the polysaccharide chain is broken down into NAG–NAM dimers and the cell falls apart. Lysozyme will also hydrolyse chitin, which is the hard cell-wall material of crustaceans, and consists of a polymer made up from NAG units alone. Thus, the additional side chains of NAM do not appear essential for lysozyme activity, though they have some effect because cleavage of the cell wall always takes place after an NAM, not after an NAG unit. However, the acetyl groups are essential for lysozyme activity. The reasons for this sort of specificity are now beginning to emerge from structural studies of lysozyme by itself and in association with analogues of its substrate.

The products of the reaction between cell-wall polysaccharide and lysozyme are disaccharides. Although the enzyme will not break them down further, it can bind disaccharides, so that when lysozyme is acting on a suspension of cell-wall material, the disaccharides produced in the reaction tend to compete with the substrate for the binding site; a high concentration of disaccharide (or, in fact, mono- or trisaccharide) tends to inhibit the activity of the enzyme. Conversely, the fact that competitive inhibition occurs suggests that the mono-, di- and trisaccharides occupy the same binding sites as the natural substrate. Whereas it is difficult to work with the substrate, it is easy to study the effects of these small sugar molecules.

The first experiment tried was to diffuse the monosaccharide NAG into lysozyme crystals in the same way as had been done with the heavy-atom compounds. This caused changes in the amplitudes of the diffraction spectra, but no disturbing changes in the crystal packing. Electron-density maps of the enzyme containing NAG showed high electron densities within the lysozyme cleft suggesting that the cleft is the active region of the enzyme. Further experiments have revealed a pattern of binding sites within the cleft. Most of the results to date have come from low-resolution maps which, of course, indicate only the general positions of the sites. However, 2 Å resolution maps have now

been worked out for NAG and the trisaccharide tri-NAG, and the features of the protein that are responsible for its activity are becoming clear.

The three sugar rings of tri-NAG run along the cleft. The ring which possesses the most highly specific interaction has its *N*-acetyl amino side chain pointing into the cleft where its hydrophobic terminal methyl group fits into a small hydrophobic pocket. Further, its NH and CO groups can form hydrogen bonds to main-chain groups on opposite sides of the cleft. Other groups of the ring are in favourable positions for forming hydrogen bonds with the enzyme. The other two rings are also in favourable situations for forming hydrogen bonds and the plane of the middle ring is parallel to the plane of one of the ring-shaped amino acid side chains, tryptophan. This parallel arrangement of rings is energetically favourable, and it is striking that the map shows a slight movement of the enzyme chain, on the more flexible left-hand side, so as to bring the rings more closely together, slightly narrowing the cleft. Of the various interactions between inhibitor and enzyme, that involving the acetyl group appears to be the most highly specific and explains why the presence of this group is necessary for effective enzyme activity.

A longer polysaccharide than tri-NAG, such as the actual cell-wall substrate, would be expected to extend further along the cleft. The positions that would be occupied by the additional units can only be inferred at present, although there is some direct evidence of a further site from low-resolution studies. Obviously, it would be surprising if the type of experiment carried out so far showed an inhibitor molecule bridging across the catalytic site, because any molecule that did so ought to be cleaved, so that at most only the part to one side of the site should remain attached to the enzyme. The catalytic site is probably therefore to be found at one end of the tri-NAG position, or one or more units beyond. It is unlikely to be at or beyond the top end, because this is right at the top of the cleft.

Further information can be sought by testing how the cell-wall substrate, consisting of alternating NAG and NAM units, would fit into the cleft. NAM has an additional lactyl side chain next to the *N*-acetyl group, and it is immediately apparent that NAM could not fit into site C. It could fit into site B, where the

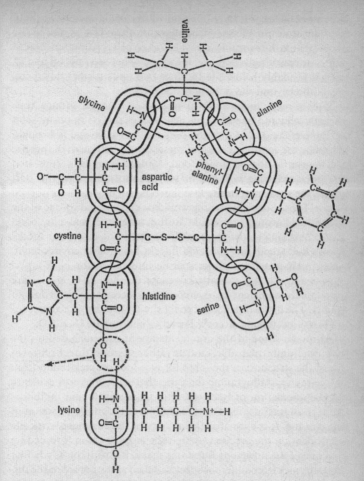

Figure 4 This hypothetical protein chain shows bonds which hold the constituent amino acids together. In the lysozyme molecule the most obvious stabilizing influence is provided by the amino acid cystine, seen here in the centre of the diagram, in which two side chains can come together to form a disulphide bridge; there are four such bridges in lysozyme. Another stabilizing force is provided by hydrogen bonds. These are relatively weak bonds that are formed, for instance, between NH or OH groups and CO groups; such groups occur in each unit of the main chain of lysozyme and in many side chains. At the bottom left is shown how an additional amino acid links to the protein chain with the removal of a molecule of water

N-acetyl and lactyl groups point outwards from the cleft. The cell-wall substrate must therefore be aligned with an NAM unit in site B. Now it is known that cleavage takes place below an NAM unit; for reasons given above, it cannot be between sites B and C, so it is probably below site D, since two units further down still would be beyond the bottom of the cleft.

Interestingly, there are two negatively charged groups (one aspartic acid and one glutamic acid side chain) on opposite sides of the cleft and it is thought possible that cleavage is brought about by the combined effect of these groups, although the precise mechanism is not yet understood. Molecular models show that the polysaccharide will not fit very comfortably into the cleft unless the sugar ring in site D is flattened slightly from its normal 'chair' form. Such a flattening would be encouraged by one of the nearby acid groups and would, at the same time, make the bond to the next ring below more easily broken. The molecular model shows that sugar ring E would have a favourable environment, again perhaps with a rather specific interaction for its *N*-acetyl side chain, so that the attractiveness of sites A–C and E are presumably sufficient to overcome the necessary distortion of ring D. The distortion needed to fill site D would also account for tri-NAG occupying sites A, B and C and not B, C and D.

Although some of the arguments are at present based on the construction of plausible models rather than on direct observations, the structure of the inhibitor enzyme complex found from the X-ray crystallographic data together with chemical evidence on the interaction of lysozyme with its substrate and inhibitors seem to suggest a coherent explanation of the mode of action of the enzyme. It is clear that the activity of the enzyme is critically dependent on the precise way that the enzyme chain is folded up and one of the intriguing questions that arise from this is whether in lysozymes from different species, which show considerable differences in overall amino acid composition, those side chains that come together to form the active region are invariably the same.

Part Two
Processes and their Control in the Organism

An organism is more than a collection of differentiated cells. Each cell has its own specialized functions to perform but this must be done in an integrated manner which is achieved by internal signalling systems – the endocrine and nervous systems. The papers in this section are concerned with the way that these systems operate to regulate the life of the organism so that it functions in an efficient economic manner. If, for example, too much of a particular hormone is being secreted, this state of over-production must be stopped for the organism to remain healthy. The excess of hormone itself initiates control mechanisms which slow down or stop its secretion. This 'feedback' control is typical of many living systems (it was described at the cellular level in Part One) and also of many man-made devices such as a fish tank with a heater controlled by a thermostat. The way such control operates is the subject of the extracts from the book by Bayliss, and the principles involved are applied to the control of salt and water concentrations in mammals in Reading 11 by Phillips and Bellamy. Here the connection between the structure of an organism, and the physics of its operation is well illustrated.

Marshall describes the nervous and endocrine control of reproductive cycles in birds and shows how climatic conditions and the behaviour of other organisms can affect this control. Wilkins in Reading 13 raises many problems associated with the control of internal rhythms, and shows that although there is quite a large body of information about them we are in no position to say how they arise, nor how they are controlled. However, by many interesting and rigorous experiments, the factors involved are being gradually sorted out. Many animal

species (including man) have well-regulated biological clocks which keep time remarkably accurately. These rhythms obviously have great importance, for instance in migration, where direction is determined in relation to the sun or the stars. Their orientation and altitude in relation to an observer on Earth is constantly changing, and allowance must be made for these apparent movements in order to maintain a constant direction. Navigators do this with the aid of chronometers, but birds have this function built in. For many organisms the biological clock can be reset and a number of experiments have determined the location of the time-keeping parts of many animals.

Perhaps the best-known system of internal coordination is the nervous system. Recently, a great deal of exciting work has been done which goes a long way toward explaining how nerves 'work', and the paper by Noble summarizes the results of a great deal of this research. The subject of the paper is the way in which the heart tissue is able to generate its own beat without external stimulation, but the article also provides a useful review of the physics and chemistry of all excitable tissues. This paper shows how 'models', in this case a theoretical model not an actual structure, can be used for advancing a science. Here a mathematical model of the expected results of an experiment was constructed based upon certain theoretical hypotheses. When the results of the experiments became available they agreed fairly well with the model. This lends support to the idea that the actual situation agrees with the theoretical structure put forward in the first place, but is in no way a 'proof' of the idea.

As Marshall's paper showed, not all behaviour is under automatic internal control. Most organisms can adjust aspects of their behaviour in the light of previous experience, i.e. they can learn. Van Bergeijk's study of a conditioned response in bullfrogs has many interesting aspects. One is the way several hypotheses are put forward to explain the behaviour of the frogs (e.g. it could have been that they 'remembered' the sequence of five working days followed by two days off at the weekend, or it could have been a response to the lights being

switched on, or it might have been due to the general noise about the laboratory on working days which acted as a stimulus for the frogs to start gathering at the feeding point). This is the point in a scientific investigation where imagination is called for. In order to form a hypothesis, a wealth of ideas is required. Then each of the hypotheses must be tested, and here again imagination is needed. Van Bergeijk goes on to test each of his hypotheses, with the aid of a computer to analyse the vast amount of data. The results are ingeniously presented in the form of contour maps representing the density of frogs.

The ability to learn probably depends partly upon genetically determined potential, and partly on the degree of development which this potential undergoes as a result of interaction with the environment. The importance of the genetic endowment is shown by the studies of the effect on man's behaviour of abnormalities in the number of chromosomes as reported by Court Brown. These cases raise moral problems: to what extent is a person with a chromosomal abnormality of this type responsible for his actions? The defence of 'abnormal chromosomes' has already been pleaded in court, and has raised some fascinating legal problems.

One of the most valuable functions of a scientific theory is that it acts as a stimulus to other research workers to design experiments to confirm it, or to disprove it. The effect of the original idea of applying the 'counter-current principle' by van Dam can be seen in both Readings 17 and 18. Reading 18 is by Hughes, who has spent a great deal of time studying the way in which fish extract oxygen from water. The neat experimental report by Hazelhoff and Evenhuis illustrates the importance of an arrangement found in many living and mechanical systems which involve the interchange of material between two fluids. The 'counter-current principle', in the case of the gills of fishes, is the arrangement of the flow of blood in the gill in the opposite direction to the water flow. This ensures that the concentration of oxygen in the water is high opposite the point in the gill where its concentration in the blood is also high, and low in the water where it is low in the gill. This ensures that oxygen uptake by the blood is always efficient. It

is useful to work out why the efficiency of the gill is reduced by reversal of the direction of water flow, as is demonstrated by the results of Hazelhoff and Evenhuis.

In this section, the level of organization is that where the cellular factory must be regulated so that its products fit in with the 'national plan' as it were. The next section will be devoted to the relationships between the organisms in a community.

10 L. E. Bayliss

Living Control Systems

From L. E. Bayliss, *Living Control Systems*, Edinburgh University Press, 1966, pp. 2–5, 143–50.

Automatic control
Some simple examples of automatic control

To begin with, let us describe a few devices which may properly be called automatic control systems. One of the earliest of these is the governor which Watt (1736–1819) put on his steam engines. The amount of steam which is fed to the engine must be made to depend on the load to be overcome; if too little, the engine stalls, if too much, it races and perhaps damages itself. A throttle valve, accordingly, is placed in the steam-supply pipe. The amount of steam flowing to the engine will depend, also, on the pressure in the boiler, and this will vary with the diligence of the stoker, as well as on the amount of steam needed by the engine to drive the load. The governor, therefore, has to take into account both the load and the steam pressure. It does this, not by attempting to measure both directly, but by detecting changes in the speed of the engine, opening the throttle when the speed falls and closing it when the speed rises; counteracting, in other words, any departure or 'misalignment' from the desired speed or 'set point'.

Another example, more familiar but much cruder, is the ball-valve in the domestic water tank. This detects changes in the level of water in the tank, admitting water from the mains whenever the level falls as a result of water being drawn off through the taps: the flow from the mains is thus kept equal to the total flow from all the taps, each of which may be opened or shut in an unpredictable way. But suppose that it was the inflow to the tank that was adjusted arbitrarily, and not the outflow from it: suppose, for example, that the bath taps are turned on without inserting the plug in the waste pipe. Within limits set by the

height of the bath, the level of water will rise until the outflow down the waste pipe is the same as the inflow from the taps. Again inflow and outflow become equal, but the system cannot properly be called a control system. The outflow is not regulated or adjusted so as to be equal to the inflow, but happens to become so owing to the inherent properties of the system: most automatic control systems work in conditions where there is no such tendency towards stabilization – indeed their chief value is in conditions where the inherent tendency is in the other way, towards a 'run-away' to some extreme state, as in an engine without any form of governor.

So far we have assumed that the fluctuations in demand that the control system has to cope with are random, without any particular pattern. But in some of the most valuable kinds of control system this is not so: they are given definite instructions which they have to act on precisely, and they do this in the same way as those already described, by comparing the results of their activities (their 'output' as it is called) with the instructions given them (their 'input') and ensuring that the difference between them (the misalignment or 'error') is zero, or as close to zero as possible. (The terms 'input' and 'output' of a control system are sometimes confusing: the input to the ball-valve of the water tank is the water level, lowered by an outflow of water; the output is the movement of the valve which controls the inflow of water.) Such a control system is called a 'follow-up servo-system', and an example is to be found in an automatic machine tool, which is fabricating parts of a particular size and shape from, say, a bar of steel. Ordinarily, the operator of the machine controls the movement of the cutting tools by hand, checking the dimensions of the part being made from time to time against those on a drawing provided. In an automatic machine, the instructions on the drawing are fed to servo-systems which control motors, advancing or withdrawing the cutting tools until the shape and size of the part correspond to those on the instructions – i.e. until the misalignment is zero. A similar kind of system can easily be imagined as controlling the whole operations of the factory, replacing the inspectorate, for example, which examines the product, rejects that which is incorrect, and issues instructions which remove the source of the rejects.

The study of living control systems may be said to begin with the work and thought of the great French physiologist Claude Bernard (1813–78). In the summer of 1870 he gave a course of lectures in the Natural History Museum of Paris; these were published in 1878, just after his death, under the title *Lessons on the Phenomena of Life Common to Animals and Vegetables*. The second lesson (the third after an 'introductory lesson') is headed: 'The Three Forms of Life': first, 'latent life' as in seeds, eggs and some kinds of animalcule which have temporarily 'fallen into a state of chemical indifference' on being dried for instance; secondly 'oscillating (or fluctuating) life', as in rather primitive animals and plants whose activity and behaviour depend greatly on the nature of their surroundings; and thirdly, the 'free and independent life', best seen in the most highly developed animals which are little affected by changes in the physical and chemical properties of their surroundings. In this lesson, also, Bernard introduces the idea that the living cells which make up an animal (and also a plant) are not in contact with the 'external' environment – the air, sea or lake in which it lives – but with an 'internal' environment consisting of a watery solution often quite different in composition from that of the outside solution. 'The constancy of the internal environment is the condition that life should be free and independent.' This constancy, however, does not mean that the animal is shut off from its surroundings and unaffected by them. 'So far from the higher animal being indifferent to the external world, it is on the contrary in a precise and informed (*étroite et savante* are the words actually used) relation with it, in such a way that its equilibrium results from a continuous and delicate compensation, established as by the most sensitive of balances.' This passage might be paraphrased in more modern terminology as: 'The higher animals possess control systems which adjust the interactions and exchanges with their surroundings in such a way that the physical state and chemical composition of the internal environment remains constant.'

Bernard's ideas were taken up and developed by the English physiologist Joseph Barcroft (1872–1947) in a series of lectures

delivered at Harvard University, USA, in 1929 and published in 1934 under the title *Features in the Architecture of Physiological Function*. In 1929, also, the American physiologist W. B. Cannon (1871–1945) introduced the word 'homeostasis' (from the Greek, meaning literally 'similar standing') to describe, generally, all the processes concerned in controlling the physical and chemical properties of the internal environment of an animal. He developed and extended his ideas in a book *The Wisdom of the Body*, published in 1932. The titles of these books emphasize the fact that control of the internal environment is essential to the whole design and organization of an animal, ensuring that its component parts work in cooperation with one another.

The extent to which life is entirely 'free and independent' will depend on the precision of the control systems concerned. There are likely to be many of these, so that homeostasis may be more perfect in some respects than in others. There is no hard and fast dividing line between Bernard's second and third forms of life: there are animals whose life is 'free and independent' in respect, say, of the salt concentration of their body fluids, but 'fluctuating' in respect of their temperatures. In studying homeostasis in any of its various aspects, our task is to discover which of the various activities, both internal and external, of the animal or plant are controlled – i.e. the nature of the input and output of each of the control systems concerned; how the misalignment between them is detected and corrected; and how accurately the system works. This is done chiefly by applying the methods of experimental physiology, together with some of those of biophysics and biochemistry. But a proper interpretation of the results requires, also, some knowledge of the fundamental theory of control systems.

Study of the growth and development of animals and plants into the 'proper' size, shape and general behaviour is based on the methods of experimental embryology and biochemical genetics more than on those of experimental physiology. These have existed only for some thirty to fifty years, while the idea that automatic control systems of the 'follow-up' type are in action is much more recent still. Here we must discover how the 'information' – as on a drawing or blue-print – is conveyed to the

control systems, as well as the mode of action of the systems themselves.

[. . .]

Control of size and shape

By the size and shape of an animal or plant, we ordinarily mean its general external appearance, determined by the nature and relative proportions of its various parts. These parts – not only those visible from the outside, but also the internal organs – have their own characteristic sizes and shapes, and each consists of an assembly of cells characteristic of that part or organ. (Some of the 'parts' into which the whole animal or plant is divided on superficial observation may contain assemblies of several different kinds of cell; legs, for example, contain both bones and muscles.) Microscopic observation shows that different kinds of cell have different sizes and shapes, particularly in respect of their internal structures or 'organelles': by electron microscopy we can see, approximately, the sizes and shapes of the larger 'macromolecules' within the cells. Smaller molecules are characterized by studying their chemical properties; this is an indirect way of observing their sizes and shapes.

Growth and differentiation

Every animal and plant begins as an egg cell, derived from the female parent, which has been 'fertilized' by a sperm cell, derived from the male parent: the fertilized egg, therefore, must contain all the 'information' and 'templates' necessary for the subsequent production of all the molecules, cells, and organs, with their characteristic sizes and shapes. The egg cell develops into the animal or plant by a process of division, first into two cells, then into four, into eight, and so on, eventually reaching billions. (There are about 10^{10} cells in the human brain, for example.) Between divisions, the cells enlarge, fresh material of the appropriate kind being synthesized out of relatively simple substances present in the solution outside the cells. As the number of cells increases, different groups of cells become 'differentiated' from one another, synthesizing different kinds of

material and taking up the special compositions and shapes appropriate to the various organs and parts of the whole animal or plant.

Study of the way in which recognizable 'characters' (sizes and shapes) are inherited from one generation to the next, showed that the 'information' about these characters resides in structural units called 'genes'. (These studies were begun by Mendel, abbot of Brünn, now Brno in Czechoslovakia, who published in 1866.) These genes are ordinarily invisible: but shortly before a cell starts the rather elaborate process of division, its genes join up into a number of threads, called 'chromosomes', which can be seen, after suitable treatment, under the microscope. In the course of division, each chromosome splits longitudinally into two parts, one part going into each of the 'daughter' cells: each of these cells, therefore, contains the same number of chromosomes as the parent cell and a complete set of genes. Every cell into which the egg divides should, therefore, be capable of developing into a complete animal or plant of the proper kind: and this is so, up to the stage at which the cells begin to become differentiated. The activities of some of the genes, presumably, then become depressed, and that of others enhanced, or perhaps modified. How this occurs is unknown, although there are some suggestive possibilities.

[. . .]

Servo-control of cellular differentiation

We may now go on to discuss how the properties of the working templates in the cytoplasm are modified during differentiation of the cell. In the first place, a fertilized egg is not homogeneous, and has axes of symmetry which are determined during its formation and 'maturation'. Many kinds of egg contain 'yolk' – a source of nutrient materials during subsequent development – which occupies part of the cell only and produces very obvious asymmetry. (The most familiar kinds of egg, those of birds, consist almost entirely of yolk.) But even when the egg cell looks spherical and homogeneous under the microscope, it can be shown to contain long-chain molecules which are orientated, with respect to the axes of symmetry, in different directions in

different parts. Thus when the cell divides, although the nuclei of the daughter cells will be identical, the cytoplasms will probably be different: these differences may affect the working templates, and possibly also, the master templates indirectly. The whole process is very complicated – much more so than the simplified account given here might suggest – and much of it is still very obscure. It would be beyond the scope of this book to go into this aspect of the problem any further, since there is another aspect in which control systems are involved.

When the protein assembled on a working template forms part of an enzyme, a 'motor' is set into action, in the sense that the quantity of enzyme formed determines the *rate* at which certain chemical reactions go on – the rate, for example, at which certain constituents of the cell are synthesized. If the *quantity* of these constituents is to reach, and be held at, some set point (as in fact occurs) there must be some feedback arrangement which stops the 'motor' when the set point has been reached. Experimental investigations have shown that such a feedback does exist. The investigations have been made chiefly on micro-organisms – bacteria and yeasts, for example – since they will live and grow when supplied with quite simple nutrient materials, e.g. glucose, ammonium salts, with a few other salts in low concentration, and water and oxygen: but similar results have also been obtained when cells of the 'higher' animals are used, and there is no reason to suppose that the conclusions do not apply quite generally. The internal composition of these micro-organisms is not essentially different from that of any other kind of cell. They contain, for example, proteins and nucleic acids, so that they must possess enzyme systems which are capable of synthesizing all the necessary amino acids and nucleotides from the glucose and ammonium salts presented to them. But if, say, some particular amino acid is added to the medium in which they are growing, so that they do not need to synthesize it, they stop doing so.

It has been found that there are two ways in which this negative feedback is accomplished. First, there is 'end-product inhibition'. The synthesis of the amino acid (to continue with this example) takes place by a series of chemical reactions which form a sequence, or chain, the product of one reaction being the raw

material of the next. When the concentration of the final product of the whole sequence reaches a certain value, the catalytic activity of one of the enzymes operating the chain is blocked – not, as a rule, that of the enzyme concerned in the formation of the final product itself, but one earlier in the sequence. The substance whose reaction is catalysed by the enzyme inhibited is usually quite different chemically from the inhibitor (the final product of the sequence); this raises biochemical difficulties as to how the block is produced.

Second, there is 'repression'. The accumulation of the final product in the cell blocks the working template responsible for the production of one of the enzymes necessary for its synthesis: it has been suggested that the enzyme protein molecule is no longer able to become detached from the RNA template. During the subsequent growth and division of the cell, any enzyme remaining becomes progressively diluted. Repression is clearly a more 'efficient' type of feedback than is end-product inhibition, since the activities of the cell are not wasted in producing useless enzymes. But it is slower in its action, since the cell may have to divide quite a large number of times before the enzyme molecules already synthesized have disappeared. We have, however, little or no evidence as to the time relations of the processes, or, indeed, much quantitative knowledge of any kind.

In some circumstances there may be a process somewhat analogous to input feed forward. A suitable kind of microorganism is grown in a medium of known composition, and is then transferred to another medium which contains some substance not previously present. This new substance may be utilized immediately, and presumably the necessary enzymes were already present, though not used. On the other hand, there may be an 'induction' period of several hours before the new substance begins to be utilized; presumably the new enzymes are being produced during this period, in response to the presence of the new substance.

It must be admitted that the precise relevance of enzyme repression, induction and end-product inhibition to the problem of cellular differentiation is by no means obvious. One has to suppose that there are other influences at work determining the 'sensitivity' of a particular cell to repression or induction, for

example: there must be a very considerable 'amplification' of the original asymmetry of the fertilized egg. Many suggestions have been made as to how this may occur, but nothing much is really known.

Control of the parts by the whole

Eventually, the processes of cell division and differentiation result in the production of a complete animal or plant. It is usually small at first, and lacking some of its parts (functional reproductive organs, for example): as it grows, its various parts are somehow constrained to grow at various rates such that they remain at, or take on, an appropriate size in relation to that of the whole animal or plant; it has a characteristic shape. This regulation is brought about by hormones secreted by certain cells in controlling 'centres'. In mammals, for example, the chief 'centre' is in the anterior pituitary body, attached to the base of the brain, though not part of it. If the rate of secretion of the 'growth hormone' is unduly small, the animal becomes a dwarf; if it is excessively large, the animal becomes a giant. In addition, there are other specific hormones which control the rate of growth of certain parts of the animal only. Secretion of some of these is controlled by 'trophic' hormones secreted by the anterior pituitary body, and is self-regulating: an increase in the concentration of the specific hormone suppresses the secretion of the trophic hormone and so stops its own secretion. There must, presumably, be some feedback from the 'target organs' on which the specific hormones act, determined by their size: how this works is not known. A less complicated example of the control of shape is to be found in many kinds of plant. The topmost, or 'leader' shoot releases a hormone which reduces the rate of growth of the side shoots; the whole plant thus takes on the form of a central stem surrounded by branches whose length is related to that of the central stem.

Although the nature of the feedback from the parts to the controlling 'centre' or 'centres' is largely unknown, its existence is strongly suggested by the facts that wounds heal up and that most plants and some 'lower' animals can regenerate whole parts that have been lost. A newt which loses its tail will grow a new one: a crab caught by a leg will detach it and then grow a

new one; a whole plant may grow out of a detached shoot ('cutting') or a small piece of root. Clearly the cells close to the wound regain their ability to divide rapidly and to become differentiated appropriately – an ability which has been lost, or suppressed, by the presence of a complete tail or leg, for example.

In this final section, we have been discussing highly complicated activities of living organisms in which there can be little doubt that control systems are in operation. But we are, as yet, far from having enough knowledge to be able to apply control-system theory with any hope of useful results. To know that it is there, however, may help to direct research which will provide that knowledge.

11 J. Phillips and D. Bellamy

The Body's Control of Salt and Water

J. Phillips and D. Bellamy, 'The body's control of salt and water',
New Scientist, 8 March 1962, pp. 572–5.

On a hot day the ambient temperature in the workshops of steel
mills and other heavy industries may rise well above 30 °C. It has
long been known that in these circumstances the drinking of tap
water to replace the large volumes of body water lost in per-
spiration causes severe muscular pain. This condition, known as
stoker's or miner's cramp, quickly disappears if plain drinking
water is replaced by a weak salt solution. As the regulation of the
salt and water balance of the body is of vital importance to all
forms of vertebrate life, the underlying mechanism deserves close
study.

Each cell of the vertebrate body is built around a complex
framework of large molecules set in a watery mixture containing
approximately 1 per cent of potassium salts. The cells are bathed
by a sodium chloride solution, derived from blood, with the same
osmotic pressure as the watery fluid inside the cell. The two
solutions are separated by a membrane so thin that it can be resolved only by magnifying the cell over 2000 times. This membrane allows the rapid passage of both sodium and potassium
between the two fluids, and normally there is a constant movement of these ions in both directions. If a solution of a potassium
salt is separated from that of a sodium salt by an artificial
membrane, sodium diffuses into the potassium solution and vice
versa, until an equilibrium point is reached at which the sodium
and potassium concentrations are equal on each side of the
membrane. Since there is always a considerable gradient for
sodium diffusion into the cell, a mechanism has been postulated
whereby sodium ions are forced out of the cell as fast as they
enter through the cell membrane from the external fluid. A large
number of the negatively charged molecules in the cell, mainly

proteins, cannot pass through, and the retention of these substances is said to account for the high potassium content of the cell; that is, potassium ions are attracted into the cell by the free negative charges. The postulated mechanism for the continuous extrusion of sodium from the cell has become known as the 'sodium pump', an analogy derived from the idea of a system which 'lifts' sodium from a low concentration to a higher concentration; thus the 'pump' opposes the direction of sodium 'flow' by diffusion.

The origin of the difference in salt composition between cells and the external fluid is obscure. Natural waters of the Earth appear always to have contained more sodium than potassium, and in this sense the fluid which bathes cells may be regarded as an extension of the external environment of the organism. It follows that any theory which maintains that the first living cells originated following the enclosure of a portion of the early seas or rivers by the cell membrane must also explain the subsequent evolution of a sodium extrusion process and the change from a 'sodium' cell to a 'potassium' cell. There is another possibility – namely, that primitive organisms originated in some form of clay, possessing negative charges which had the capacity to bind potassium selectively; thus living matter may have always been associated with a high potassium concentration. In any case most present-day living organisms function normally only if the cells contain more potassium than sodium. Deviations from a certain concentration of either sodium or potassium interfere with the action of enzymes which maintain the chemical reactions of life, so that the salt content of the body must be controlled within narrow limits.

In mammals there are various natural processes which affect the volume and salt content of the fluid bathing the cells and hence the cellular fluid with which it is in equilibrium. One of the most important of these is the excretion of urine, which involves the loss of sodium and potassium from the body. These losses are usually made good by normal feeding processes. In the absence of adequate external sources of salt and water an obvious way of combating changes in the composition of body fluids would be to cut down on losses in the urine.

Urine formation is necessary for the removal of certain waste

products, mainly urea, from the body and takes place in the kidneys by the filtering of blood through the walls of blood vessels into narrow tubes (tubules). The salt content of the filtrate is initially almost indistinguishable from that of the plasma. Usually over 90 per cent of the filtrate is absorbed back into the blood as the fluid passes through the tubules. The remaining solution drains into the bladder and is voided as urine. On the average, man produces two and three-quarter pints of urine per day containing a quarter of an ounce of sodium and one-twelfth of an ounce of potassium. Under conditions where the continued loss of this quantity of water and salt cannot be rectified by eating and drinking, a greater proportion of salt and water in the blood filtrate is reabsorbed by the cells lining the tubules of the kidney.

The kidney reabsorbs sodium by means of 'sodium pumps' located in the cells lining the tubules. These transfer sodium from the fluid in the tubules to maintain a higher concentration in the cells. Water is then passively reabsorbed, passing from the tubule to the cells as a consequence of the resulting osmotic gradient. Thus, water reabsorption is dependent upon sodium reabsorption and it appears that the greater the difference in sodium concentration that the pumps can maintain across the cell membranes of the tubule, the greater the rate of water movement through the membranes and the greater the conserving power of the kidney; so less sodium and water is lost in urine.

The automatic system for varying the composition of urine is largely under the control of hormones, produced by the posterior lobe of the pituitary gland and the adrenal glands, which are released in response to changes in the salt content of the body. Urine voided by humans on a normal diet contains a greater ratio of sodium to water than the fluid bathing the cells. Therefore, fasting plus a shortage of drinking water, together with a continued excretion of a normal urine, would bring about a fall in the sodium concentration of the blood and a decrease in blood volume. Alterations in the properties of the blood are registered by specialized cells which are able to detect variations in its sodium concentration, osmotic pressure and mechanical pressure. In response to sodium and water loss by the body, these receptors bring about a release of antidiuretic hormone from the posterior pituitary gland and of other hormones from the adrenal glands.

J. Phillips and D. Bellamy   133

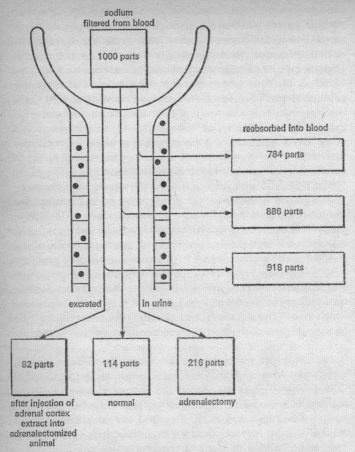

Figure 1 Sodium reabsorption in the kidney. In the kidney, blood is filtered under pressure through the walls of a capillary network embedded in a cup-shaped depression called the glomerulus and passes down the tubules to the bladder. The blood cells and proteins are retained in the blood so that the filtrate contains mainly the inorganic constituents. Most of the sodium in the filtrate is reabsorbed back into the blood through the cells lining the tubules and this process is under the control of adrenal hormones

Antidiuretic hormone specifically stimulates the process by which water is reabsorbed from the kidney tubules. In an unknown manner this substance brings about a decrease in the physical barrier to the osmotic movement of water through the cell membranes lining the tubules. The administration of antidiuretic hormone to normal animals gives rise to a urine similar in composition to that formed during the deprivation of drinking water.

The active hormones of the adrenal gland are formed by the outer part (the cortex) and belong to the group of compounds called steroids. Surgical removal of the adrenal glands from experimental animals (adrenalectomy) eventually causes death. In the absence of the glands there is an unusually low rate of sodium uptake by the kidney tubule cells. Sodium reabsorption may be returned to normal by the administration of an extract of the adrenal cortex. Several steroids occur in the extract which have similar effects on sodium reabsorption. Minor variations in molecular structure, however, are associated with large differences in biological activity. The adrenal steroid with the greatest biological activity in respect of renal salt transfer is called aldosterone.

Steroids of the adrenal cortex and antidiuretic hormone from the pituitary gland are released together and act in combination on the reabsorptive mechanisms of the kidney to conserve both water and salt. However, an excess of antidiuretic hormone and aldosterone cannot bring about the complete reabsorption of the blood filtrate. That is to say, in face of continued salt and water deprivation the minimum volume of daily water loss through the kidneys is only reduced to one-third of the normal amount. Under these conditions there is also a limitation on the quantity of sodium which can be removed by the tubule cells so that reabsorption of all the filtered sodium is never achieved. In twenty-four hours the amount of urinary sodium unavoidably lost is just over one-tenth of an ounce. This sodium is dissolved in about a pint of water to give a solution with almost twice the sodium concentration of blood. Obviously if a thirsty man drank one pint of sea-water per day (which contains about two-tenths of an ounce of sodium), although the urinary loss of water would be counteracted, the sodium content of the body would rise by about

one-tenth of an ounce. This progressive accumulation of salt is the main reason why man cannot make up water losses on a diet which includes sea-water as the sole source of water.

It can be seen that man cannot survive under arid conditions without a source of fresh drinking water because of the limit on

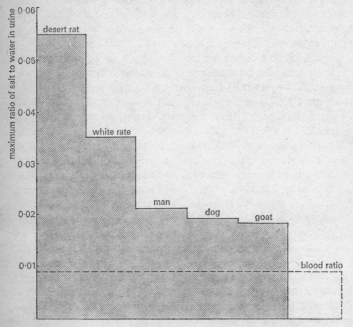

Figure 2 Salt-concentrating ability of the mammalian kidney. The given ratio of sodium to water in urine of various animals or the maximum ratios which have been obtained under experimental conditions. The ratios are related to the maximum amount of water which can be reabsorbed per sodium ion pumped out of the tubule

the maximum ratio of salt to water in urine. However, several mammals have successfully colonized the deserts of the Earth, and their success in this exacting environment can in many species be related to the greatly improved reabsorptive and concentrating power of the kidney. For example, the small rodents (desert rats) which inhabit the hot dry regions of the world

normally form minute quantities of urine. The sodium concentration of this fluid is usually about twice that of sea-water. In this way, by drinking 5 ml (about one-hundredth of a pint) of sea-water per day these animals are able to retain 50 per cent as pure water to make good extra-renal water losses. Drinking salt water is not the normal method of keeping in water balance, however, because the desert rat can satisfy its water requirements with the water contained in food and that produced by the biological oxidation of foodstuffs.

The ability of desert rats to bring about an almost complete reabsorption of the blood filtrate is associated with an unusually large quantity of antidiuretic hormone in the blood. This circumstance may well facilitate the reabsorption of a larger number of water molecules per sodium ion 'pumped' from the tubules than is possible in other mammals, a condition which is also necessary for an increase in the urinary sodium concentration. The available evidence would also suggest that the adrenal cortex has a greater capacity for aldosterone synthesis than that of other mammals. Thus an increased concentration of aldosterone appears to be partly responsible for the high sodium concentration in the tubule cells which is necessary for maximum water conservation.

Vertebrates other than mammals – for example, birds and fish – also face environmental hazards which tend to change the composition of the body fluids. The sodium concentration of the body fluids of all vertebrates, with the exception of the primitive hagfishes, is roughly similar and it would seem that similar hormones from the adrenal glands and the posterior lobe of the pituitary gland are concerned in the process opposing any changes in body fluid composition. The same adrenal hormones are found from fish to man and the antidiuretic hormones display only minor variations in structure throughout the vertebrate series. These two types of hormone, as we have seen, are concerned in mammals with kidney function and this appears to be equally true for most of the other vertebrates, but in addition non-mammalian vertebrates have supplementary excretory mechanisms which combine with the kidney to offset salt gain and water loss.

All birds possess paired organs situated above the eyes called

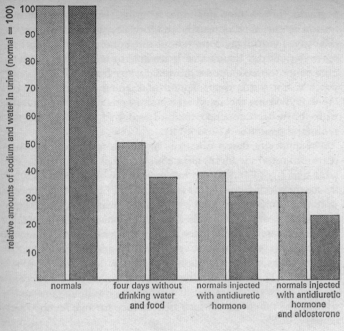

Figure 3 Relative amounts of sodium and water excreted in human urine. The average amounts of sodium and water in normal human urine are taken as 100 and the amounts excreted under various conditions are expressed as percentages of the normal quantity. It can be seen that there is a marked reduction in sodium and water loss, and an increase in urinary sodium concentration during a period without food or water. This condition may also be brought about by the injection of antidiuretic hormone into normal subjects. Aldosterone and antidiuretic hormone together are more effective still in conserving water, indicating that water reabsorption is related to the reabsorption of sodium: that is, the extra sodium reabsorbed in the presence of aldostrone brings about increased water reabsorption

nasal or salt glands. When marine birds drink sea-water, a fluid appears from the nasal region; in some birds the fluid can be seen to collect as a drop at the tip of the beak and in others the fluid is ejected through the nasal openings by a characteristic

shaking of the head or by a 'water-pistol mechanism'. This fluid has been shown to originate from the nasal glands and is essentially a sodium chloride solution which is more concentrated than sea-water. Such a secretion enables the bird to avoid sodium accumulation and survive indefinitely with sea-water as the sole source of drinking water. Following the ingestion of sea-water, birds which have been treated with adrenal hormones excrete greater amounts of fluid from the nasal gland than do untreated animals. Conversely, secretion by the nasal glands may be decreased by the removal of one adrenal gland and completely prevented when both glands are removed. It would seem, therefore, that a hormone from the adrenal gland is an essential factor in nasal-gland secretion. Supporting evidence for this idea comes from experiments in which sea-gull chicks were reared exclusively on sea-water or tap water. After six months those birds on sea-water were found to have significantly larger adrenal glands than those given fresh water to drink. This finding is in agreement with observations which reveal that in general marine birds have larger adrenal glands than terrestrial birds.

The nasal glands of the domestic duck also have the capacity to respond in a similar way to those of truly marine birds. Following the ingestion of sea-water the nasal glands of the duck are triggered into action by the rise in blood sodium concentration. It is probable that the response is indirect and is mediated by cells similar to those already indicated in mammals, which are sensitive to changes in the blood sodium concentration; these cells in turn stimulate the release of antidiuretic hormone and the adrenal hormones. Thus the stimulus of increased blood sodium results in a changed complement of hormones in the blood. The receptor cells appear to be extremely sensitive because maximum secretion by the nasal glands may be initiated by less than a 0·5 per cent change in the blood sodium concentration. The hormones, as in the mammal, also act on the kidney to facilitate the reabsorption of the blood filtrate. It appears that the formation of urine by birds drinking a concentrated sodium chloride solution would, because of the inadequate concentrating ability of the kidneys, aggravate the undesirable effects of the increased sodium content of the body fluids.

When cortisol, another steroid produced by the adrenal cortex,

is given to a bird from which the adrenals have been removed it completely restores the ability of the nasal glands to operate; further, cortisol and deoxycorticosterone, but not aldosterone, have been found to increase the amount of salt excreted by the

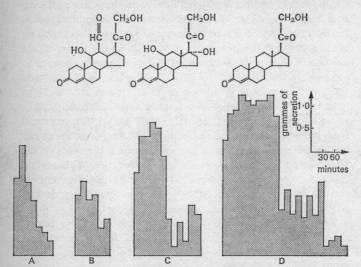

Figure 4 Secretion of salt water by the nasal gland of the domestic duck. A: a duck was given an intravenous infusion of 40 cm³ of sodium chloride solution at the same concentration as sea-water. About one-fifth of this was secreted by the nasal gland in two hours. No urine was voided in this time. The experiment indicates that the duck can tolerate an appreciable rise in blood sodium (15 per cent increase, equivalent to four-fifths of the sodium chloride administered) before the nasal gland is activated. Thereafter less than a 0·5 per cent increase is sufficient to bring about secretion. B, C and D: nasal-gland secretion following the injection of the same amount of salt solution into ducks treated with various adrenal steroids (B = aldosterone; C = cortisol; D = deoxycorticosterone). Slight alterations in the chemical structure of the steroids bring about large differences in their ability to stimulate nasal secretion

nasal glands of normal ducks given sea-water. Thus it would appear that a cortisol-like hormone is one essential factor in the initiation and regulation of nasal-gland secretion of normal ducks given sea-water. Marine birds, therefore, possess mechanisms of

hormonal control, which under conditions of excess salt intake and water deprivation restrict kidney function to a minimum and at the same time bring into action a more economical method of excreting salt and conserving water.

Birds are not unique in possessing tissues which at times may assume a greater importance than the kidneys in regulating salt and water balance. Fish living in fresh water or sea-water have, respectively, problems of water elimination and conservation. In sea-water, which is nearly twice as concentrated as body fluids, the needs of fish are similar to those of marine birds. That is, a salt-excreting mechanism is required because sodium diffuses into the body from the external environment. There is also a water loss by osmosis through the body surfaces. Although the urine volume is lower in marine fish than in freshwater species, the restriction of water loss is not sufficient to keep the animal in water balance and marine fish must drink sea-water to make up for the deficit. The latter process also requires a mechanism for the removal of surplus salt (sodium pump) and evidence points to this being situated in the gills.

For fish in a freshwater environment the converse situation applies because the body fluids are about thirty times more concentrated than the external environments. Water now enters the animal by osmosis and is excreted by the kidneys. The renal excretion of water also entails the unavoidable loss of salt (see mammal). Although some of the sodium deficit may be made up by the sodium contained in the food, an additional mechanism is required by freshwater fish to maintain the balance. Certain cells in the gills appear to have a function of 'pumping' sodium into the animal from the external environment.

Some fish, such as the eel and salmon, can live in both sea-water and fresh water. While it is clear that in a marine environment these forms drink sea-water, little is known about the physiological changes which enable them to move freely between the two environments. We may tentatively conclude that some of the adjustments are brought about by changes in the secretion of hormones. In particular, adrenal and antidiuretic hormones appear to be involved in ways not too dissimilar from those in which they aid in the survival of higher vertebrates. In other words, the success of a salmon in migrating into fresh water may

depend on its ability to switch from a sea-water type of hormonal control, with high antidiuretic hormone and perhaps low aldosterone, to the freshwater type of control, a low level of antidiuretic hormone with high aldosterone, as it swims through the estuarine waters. Besides its renal effect aldosterone is known to bring about a decrease in the outward diffusion of sodium through the body surfaces, and in this manner aldosterone secretion appears to be necessary in the freshwater fish to prevent undue sodium loss through the gills. The whole process of adaptation to the new environment seems to be very complex and the controlling mechanisms are by no means clear. Not the least of the present problems is how the 'sodium pumps' in the gills reverse direction from ejecting sodium in sea-water to taking in sodium in fresh water.

In conclusion, it appears that the hormones of the adrenal cortex and posterior pituitary gland, which have a similar chemical structure in all vertebrates, are universally concerned in regulating the salt and water balance of the body. These hormones appear to modify the properties of cell membranes in a way that secures an alteration in the rate at which salt and water pass between the cell contents and the fluid in contact with the membrane. The main sites of hormone action are in specialized 'target' organs which, although they may differ greatly in gross appearance from one species to another, have in common the role of maintaining within well-defined limits the salt composition of the blood and hence the salt content of the cell.

12 A. J. Marshall

The Environment, Cyclical Reproductive Activity and Behaviour in Birds

A. J. Marshall, 'The environment, cyclical reproductive activity and behaviour in birds', *Symposia* (*Zoological Society of London*), vol. 2, 1960, pp. 53–67.

Introduction

If we look at the world, rather than merely the European and North American bird fauna, we will see very clearly the operation, and the utility, of an internal rhythm of reproduction. It is, in fact, the most important single, and only universal, factor in the regulation of breeding seasons, including migration. By internal rhythm I mean the metrical succession of correlated events that are variously under endogenous and exogenous control. At the outset, too, I must emphasize that although we have in it something akin to a clock, it must be nevertheless likened to an imprecise, chain-store variety that has to be 'corrected' by the external factors at least once during each annual cycle.

The reason why the existence of this partly endogenous rhythm is not more widely accepted is that almost all investigators have restricted their attention to temperate-zone birds, and have allowed the spectacular influence of exogenous photostimulation in many such species to obscure any clear view of the basic mechanism underlying it.

Rowan's pioneer work led to an era of photostimulation experiments which have been productive of a great deal of vital information. Most people do not realize that it was this great innovator, in fact, who was the first (as far as I can gather) to postulate the existence of an internal rhythm of reproduction. Rowan, at the height of his photostimulation work in Canada, concluded that some kind of rhythm must be operative in the regulation of the breeding cycles of equatorial species. In Africa, Moreau, with whom Rowan corresponded, came to the same conclusion. Later, F. H. A. Marshall and Baker and Baker all produced *prima facie* evidence of the operation of an endogenous

rhythm. Baker and Baker were insistent that such a rhythm could not be indefinitely self-regulatory and that external stimuli must operate from time to time to synchronize it with the cycle of the sun, i.e. with the seasons. Marshall produced additional information and suggested that, notwithstanding the vital influence of photostimulation in many species, the most positive timing and regulating (as distinct from initiating and accelerating) factors are the external stimuli in the traditional breeding grounds that operate, via the female exteroceptors, only when the environment becomes seasonally suitable for reproduction.

Photoperiodicity does not, even in most temperate-zone birds, control the *time of breeding*. It does however, initiate the 'spring' (often late winter) sexual processes in very many temperate-zone species at a time sufficiently early to allow an adequate period for the acquisition of territory, display and the other hormone-induced processes necessary, along with additional external stimuli, for successful reproduction. In probably hundreds of species living in tropical, and periodically dry (not necessarily desert) regions, light has no influence whatever. We shall see that *light is important only in species for which it is important that light shall be important*.

The avian sexual cycle

The partly endogenous and partly exogenous rhythm of reproduction, involving successive phases of post-nuptial *regeneration* (reflected in a sudden loss and subsequent recovery of breeding function), *acceleration* (characterized by sex-hormone production and gametogenesis) and *culmination* (involving ovulation and insemination) can be considered as a sort of cog-wheel which is engaged at various times by changing environmental 'teeth' that are helpful, or otherwise, to reproductive success and to the existence of the species.

The internal rhythm is the primary seasonal *initiator*. As young birds mature, so does their neuro-endocrine apparatus. Young budgerigars (*Melopsittacus undulatus*) and zebra finches (*Peophila castanotus*) will produce bunches of spermatozoa within sixty days in almost total darkness if given adequate warmth, food and water. These are temperate-zone birds which live in dry, but not desert, areas. In the zebra finch, and numerous other species,

reproduction can occur within six months of hatching. Likewise, among adults the internal rhythm is the seasonal initiator. Until the post-nuptial regeneration phase is past, the male certainly, and the female probably, is uninfluenced by the external stimuli that would cause gametogenesis at other periods of the cycle. After the spontaneous progression from regeneration to acceleration (when sex hormones are once more liberated and gametogenesis is again possible) the neuro-endocrine machinery of each individual comes under the constant influence of two sets of external factors as follows:

accelerators $\rightleftharpoons$ *inhibitors*.

These are antagonistic. The cycle hastens, slows or sometimes stops altogether, depending upon the factors currently presented to it by the changing environment. The more unstable the climate, the more obvious the truth of this becomes. For example, an accelerator of the sexual cycle of many temperate-zone birds is light. An inhibitor of the cycle of most species is cold. In an unusually cold spring the effects of light are nullified and so reproduction is delayed. Among water-birds and many others a frequent inhibitor is the lack of a safe and traditional nesting site. If the seasonal appearance of such a site is delayed sufficiently long in an habitually single-brooded species, internal changes occur that lead to metamorphosis of the testis and so reproduction is frustrated for that particular season. Other accelerators are the innate behavioural reactions induced in each sex by their gradually changed endocrinology, as well as by the presence of the displaying mate. Yet ovulation, even after the first egg is laid, may be inhibited by sudden cold, fear or other traditional inhibitor.

Regeneration phase

This comes immediately after reproduction in single-brooded species and after the final ovulation of the season in plural-brooded ones. It varies in duration between species. It is a period in which the neuro-endocrine apparatus does not respond to photostimulation, hence the use of the term 'refractory period' by Bissonette and Wadlund. It is accompanied by the almost complete metamorphosis of the testes. After the seasonal exhaustion, or near exhaustion, of the interstitium, the Sertoli and germ-

cell cytoplasm are converted into a mass of cholesterol-positive lipid material which appears to give rise to progestins. The same effects can be induced by the injection of prolactin, or by the removal of the anterior lobe of the pituitary gland. Automatically, and even after the removal of the hypophysis, the interstitium rhythmically regenerates and the tubule lipids disappear. The rate of disappearance varies between species. Yet there is no suggestion that the interstitium could *secrete* in the absence of the hypophysis.

Although the testis is thrown out of spermatogenetic productivity during the regeneration phase, and there is evidence that the ovary also becomes 'refractory', these events are governed from elsewhere. Most workers suggest that it is the anterior pituitary that is the seat of seasonal negativity. There is no anatomical evidence that this is so – yet there is both experimental and histological evidence that the adenohypophysis is at least involved. The testes of Arctic and desert non-breeders sometimes metamorphose before they reach full spermatogenesis. That is probably due to effects, exerted by the unsatisfactory state of the environment, on the central nervous system and, it follows, on the anterior lobe which then ceases to produce gonadotrophins.

The evidence customarily used to implicate the anterior lobe could apply equally well to the hypothalamus, or any other part of the central nervous system that periodically activates the adenohypophysis. We know that single-brooded species regularly become 'refractory' after one nesting, and plural-brooded birds only after several, and this may suggest that the factors controlling the onset of this regenerative period of sexual negativity are neural ones. Ultimately these may be dictated by external events including, particularly, behavioural reactions between the pair.

During the regeneration phase sexual behaviour is nil, or almost so. Many normally solitary species tend to flock, often in company with other species. The post-nuptial moult begins, and proceeds strongly. The duration of the period in the wild bird appears to be stable within the species. As a result of photostimulation work, it can now be predicted when a given spring-breeding species will end its regeneration phase, i.e. cease to be 'refractory'. Most North-temperate-zone species end their regeneration period between August and November. Some believe that a

period of short days or long nights of winter are needed before the regeneration period will end. We will see that in many species no such influence is necessary.

Acceleration phase

This follows the regeneration phase automatically. The bird has now reached a physiological condition in which its neuro-endocrine apparatus is susceptible to external stimuli. Morley and Marshall have provided numerous references that many British birds, and others elsewhere, undergo autumnal sexuality. This follows the renewed secretion of gonadotrophins which, in turn, reactivate the gonads and cause the flow of sex hormones, the sharp or gradual disappearance of the last remnants of tubule lipids and the occasional autumnal multiplication of germ cells. This phase varies considerably in its duration between species as well as in relation to the state of the environment, including the partner and the flock. In most birds the post-nuptial moult ends early in this phase. Among Arctic and desert non-breeders, as well as birds in zoos, progression is halted by environmental influences. The acceleration phase is a period of territory selection, song and increasingly intensive sexual activity. Some species develop secondary sexual characters (e.g. the assumption of beak colours and nuptial plumes). An early start of such activities (in autumn) is probably conducive to later reproductive success. At the same time, external inhibitors (e.g. cold) will temporarily halt the process late in autumn and prevent the wastage of the reproductive potential in winter. Among many tropical species the advent of dry weather, or otherwise unfavourable external conditions, does likewise. Among other tropical species, e.g. Ascension Island sooty tern, which breeds four times every three years, and certain other birds—the absence or relative absence of such inhibitors permits the abandonment of the more or less precisely annual cycle. As other investigators turn their attention to different parts of the world, more and more species will be found to exhibit more than one complete cycle each year.

In temperate-zone birds whose acceleration phase began in late summer or early autumn, and was depressed or halted in winter, there is after the winter solstice a sharp, probably photostimulated, renewal of sexual behaviour, accompanied by gametogenesis.

The cycle now continues under the influence of spring stimuli and inhibitors until its culmination with reproduction.

Culmination phase

Now nest building is completed and insemination and ovulation occur. A special *species requirement* of end-stimulation is required before the cycle culminates in reproduction. Towards the end of the acceleration phase the testes contain massive bunches of spermatozoa. Many are free in the lumina and great numbers have already reached the seminal vesicles. The female, whose accessory sexual organs have been enlarging but which has nevertheless lagged behind the male in her gametogenetic development, is ready to begin the final swift oocyte development that will end in ovulation and egg deposition through a now grossly hypertrophied oviduct. But although the internal physiology may be now wholly prepared for reproduction, this phase, like that of acceleration, is highly susceptible to stimuli (e.g. that of the nest, behavioural interactions) and inhibitors (e.g. sudden cold, fear or nest-destruction). In many, perhaps most, species it is the end-stimuli that permit nidification, insemination and ovulation which are the real 'timers' of the sexual cycle. There is no evidence that the current photoperiod has any influence on it whatever. The period lasts until the final clutch is laid, after which individuals once more enter the phase of sexual negativity and regeneration described above.

Internal rhythm in aseasonal breeders

Great numbers of species both within the tropics and outside them do not breed according to any particular phase of the cycle of the sun. It is in such species that we see most clearly the operation of an internal rhythm of reproduction. The supreme virtue of light as a regulator is its regularity; but likewise (a point which cannot be over-emphasized) this unvarying regularity must be a force of destruction to any animal compelled to obey it in a habitat in which good conditions for the survival of the young did not accompany a particular phase of periodicity. It is a condition of their survival that many species reproduce at times when rainfall has rendered the environment propitious for the development

of their offspring. If the sexual cycle of many xerophilous (not necessarily desert) species was governed by photoperiodicity the young would be launched, more often than not, into arid conditions which would not allow them to survive.

We have seen above that at least some such species have developed a mechanism which allows them to become sexually mature at the age of sixty days. Thus, budgerigars and zebra finches will produce spermatozoa at that age even when kept in almost total darkness provided they are given adequate warmth, food and water. In aviaries both the above-mentioned species will reproduce almost all the year round under optimum conditions. It is an engrossing thought that the mechanism evolved by budgerigars around Oodnadatta is the same that permits the species to breed the year round in captivity at Stoke Poges. This aspect of drought adaptation is one of the factors that has allowed the budgerigar to become perhaps the commonest bird in both these rather different environments. In the dryer areas of Australia their vast flocks undulate over the horizon and are responsible for the specific name *undulatus*.

We have dealt in detail with only two well-known species, but Serventy and Marshall have indicated elsewhere that the effects of chance and other rainfall, coupled with temperature, are the essential breeding regulators of a great diversity of species over wide areas of Western Australia. It is obvious that not only physiological, but also behavioural adaptations must be involved in the sexual cycles of many such birds. In particular, display must be abbreviated.

In the temperate-zone satin bowerbird (*Ptilonorhynchus violaceus*) of the lush forests of eastern Australia the flashing, staccato display – which, incidentally, strongly resembles that of a Spanish flamenco dancer – may continue for up to four months. In temperate-zone species which live in dry areas of inland Australia, such prolonged splendour is impossible: species leap from sexual inactivity to full productivity in a matter of days. They must conclude their display, reach reproductive condition and produce their young while the beneficial rain-induced conditions last in their precarious environment. Thus, McGilp made the extraordinary observation that the desert bushchat (*Ashbia lovensis*)

began nesting five days after the beginning of a two-day down-pour which broke a prolonged south Australian drought. It laid eggs eleven days after the beginning of the rain.

The desert bushchat mentioned above, and many of the species investigated by Keast and Marshall, inhabit periodically drought-stricken semi-desert and depend for reproductive conditions on chance rainfall. There is also a large category of species that live in *relatively* dry areas which receive one or sometimes two rain-falls every year. While travelling in the Kimberley region of north-western Australia recently, Serventy and myself found that long before the fall of expected rains, the sexual cycle of some birds was actively in progress. In such areas the advent of the expected rainfall would bring the cycle to fruition. When, as fre-quently happens, rain does not fall, the cycle is inhibited.

The grey teal (*Anas gibberifrons*) of Australia is an arresting ex-ample of a species that has been extremely successful in its be-havioural and physiological adjustments to a capricious climate. Dependent over great areas of its range on temporary water-holes filled by chance rainfall, the grey teal has abandoned an annual cycle. Furthermore, it has reduced to a minimum the long court-ship period characteristic of the genus *Anas*. By nomadic move-ments it finds water-filled clay-pans and billabongs where its sexual cycle quickly reacts to produce young while the water, and, it follows, food lasts. Remarkably, the grey teal has become the most successful Australian duck.

Australia has long enjoyed a curious reputation for peculiar phenomena – 'songless bright birds' (an absurd European belief), trees that periodically shed their bark instead of leaves, mammals that lay eggs and so on. There is however, nothing mystical about the breeding seasons of Australian birds, and to emphasize this I will turn to another great land mass, Africa.

Here, many species – both aquatic and land birds – have like-wise adjusted their physiology and behaviour to take advantage of propitious conditions whenever they occur. For example, the small xerophilous dioch or red-billed weaver-finch *Quelea quelea* has mastered its arid environment and become one of the com-monest birds in the land. This species is so numerous in vast areas of Africa that it reaches plague proportions and today

breeding colonies are systematically blown up in an attempt to bring the pest under control.

Although *Q. quelea* retains an ancestral capacity for photostimulation, such is not influential under natural conditions. If it were, the species (the adults of which can subsist on seeds only) would reproduce at periods during which the environment lacked the materials traditionally used for nesting and the amino acids essential for the maturation of their young.

The reproductive cycle of *Q. quelea* has been experimentally studied by Disney, Lofts and Marshall in order to determine the length of the regeneration (refractory) period. We have seen that in the various spring-breeding temperate-zone species so far investigated this period lasts for several months. For how long does this temporary period of enforced sterility last (if it exists at all) in the reproductive rhythm of a species in which successful reproduction can occur only after rainfall that produces conditions propitious for breeding and the survival of young?

It was found that a post-nuptial period of sexual regeneration does in fact exist, but that it is relatively brief. Thus, after each breeding season the males, at least, progress with little delay into a condition whereby they can take advantage of propitious breeding conditions whenever these arise in the arid environment to which they, and especially their reproductive processes, are adapted. Again we see a rapid completion of courtship activity. Within a few days of the arrival of the horde on the nest-site, each male starts to build and to display vigorously in the incomplete structure. This attracts a female and copulation has been observed in and alongside the nest.

Numerous other species in many parts of the world will be found to behave like those dealt with above. I see no reason why rain-induced reproduction should not be discovered in the southwestern United States and Mexico. Certain aquatic birds on the Victoria Nyanza, finches of the genus *Munia* in Borneo, have abandoned an annual cycle. Miller has carried out important work in Colombia by the use of rings and has shown that *Zonotrichia capensis* can reproduce at any time, and that individuals breed rhythmically twice in a single year.

We have seen that both physiology and behaviour – in fact, the

entire *biounit* – can change according to the environment to which the animal adapts itself. We have seen further the remarkable phenomenon of creatures mastering harsh unpromising conditions to such a degree that they achieve a greater biomass than closely related neighbours enjoying exceedingly lush conditions.

At the same time, it has to be confessed that we know little of the details involved. As mentioned before, there is evidence that a species can ovulate within eleven days of the long-delayed fall of rain. It follows that, in those species that require rain-induced cover in which quickly to nest, the preliminary stimulus should be rain itself. As yet there appears to be only one piece of evidence that this is the case. Marshall and Disney designed an experiment that showed that first-year *Q. quelea* were influenced by the green grass that grew after rainfall in tropical Tanganyika. When the experiment was repeated with *older* birds it was found that rainfall itself appeared to exert a strong stimulus to building. That rainfall (or humidity) affects the neuro-endocrine system was indicated by old non-breeders which moulted from one breeding dress to another, without the customary intervening period of neutral plumage.

We know that once the lengthy regeneration period has ended in some species, increased light can stimulate the sexual cycle (including behaviour) which then comes under the control of various stimulating environmental factors (of which temperature is of almost universal importance) significant to the particular species. It seems almost certain that, in numerous other species, once the abbreviated regenerated period has ended, rainfall *per se* can stimulate the sexual cycle which then likewise comes under the control of a variety of significant external factors. Each kind of bird has its own innate 'species requirement'. The individual components of this complex will have to be determined species by species.

Internal rhythm in trans-equatorial migrants

We have seen that the post-nuptial period of regeneration has been foreshortened in species for which it is beneficial that reproduction should occur at any time of the year. I believe that this endogenous component of the cycle is adaptive in the other direction and that a lengthening of the period of sexual negativity may be the principal timing factor in equatorial and trans-

equatorial migrants which begin their nuptial journey under almost constant, or decreasing, day lengths. For example, some populations of yellow wagtails (*Motacilla flava*) 'winter' on the equator (or south of it) and arrive in their European breeding grounds with considerable regularity each northern spring. It seems possible that, after the automatic emergence from a prolonged regeneration period, involving sex-hormone liberation, *various* environmental factors (including particularly behavioural interactions) start the cycle anew. Thus, the behaviour of yellow wagtails 'wintering' on the equator begins to change some time before migration. With the first manifestation of gametogenetic development in each sex, a greater social cohesion becomes apparent and the birds begin to associate in pairs, no doubt as a result of the post-regeneration production of sex hormones. After a further short period of almost constant day length the birds leave for their northern nuptial grounds where their breeding season (and sexual cycle) is probably ultimately timed by the external factors that permit successful nesting and reproduction. Laurence Irving has evidence that *M. flava* arrives in Alaska (for example) with considerably enlarged testes. Eggs have been recorded within ten days of the arrival of the birds on the breeding ground.

That an internal rhythm of reproduction does in fact exist in one trans-equatorial migrant, the short-tailed shearwater (*Puffinus tenuirostris*), has been experimentally proved by Marshall and Serventy. This species undertakes annually a mighty circum-Pacific journey from southern Australia to Japan, the Aleutians, western North America and back across the Pacific to Australia with astonishing regularity. It returns to its breeding island during the same twelve days every year. Marshall and Serventy showed that the migratory flocks of this 'mutton-bird' regularly leave southern Australia in April (autumn) at a time when their principal food – euphausiid plankton – remains abundant, and when sea temperatures are still higher than at the November period of ovulation. It was concluded that the post-nuptial northern journey via Japan to the Aleutian and South Arctic contra-nuptial ('wintering') areas is probably in obedience to an inherent internal rhythm. When the millions of birds return and first make their landfall at the breeding islands in the last week in September (spring), gametogenesis is already well advanced. The testes con-

tain bunched spermatozoa. There is no doubt that both the spring sexual cycle and the southern journey are initiated under conditions of decreasing day length whilst the birds are still north of the equator. However, the astonishing regularity of their Australian land-fall and subsequent egg-laying (19–21 November and the following twelve days) suggests that the nuptial journey must be 'timed' by some astronomical constant, possibly decreasing day lengths, as appears to be the sexual cycle of certain fish and mammals. No obvious periodic changes that might operate as precise 'timing' factors were observed in the breeding area.

By varying the light regimes of experimental captives it was found that decreasing day length seems to have no effect whatever. Both experimental and control birds came into breeding condition at approximately the same period as the wild population on the completion of their annual journey. Thus, the breeding rhythm is persistent at least for one year after the animal is taken out of its normal environment and exposed to external influences (photoperiod, temperature, food and habitat) exceedingly different from normal. It must be emphasized, nevertheless, that although gametogenesis and pre-migratory fat deposition took place under these conditions at approximately the normal periods, there was in some individuals a considerable asymmetry of the seminiferous tubules and a pronounced retardation of spermatogenesis that is never encountered in the wild population. This showed that internal events were tending to drift out of phase with the external environment and that as yet undetermined external influences, possibly including behaviour, were lacking. In short, although an internal rhythm indubitably exists, it must be regulated annually if it is to remain appropriately synchronized with the seasons.

Discussion

It should be emphasized that the statements set out above perhaps represent only the opinions of myself and a group of people who have worked in my laboratory and collaborated with us in studies carried out elsewhere. It should be pointed out, too, that the more 'orthodox' views are those held by my friends Farner and Wolfson (and others) who ascribe a far greater general importance to photostimulation than do we. Wolfson in particular

argues that, even on the equator, the migration and reproduction of birds are under photoperiodic control.

We believe that a great number of species, even those existing in the temperate zones of considerable photofluctuation, are uninfluenced by photoperiodicity and that were they so influenced they would be much less successful and, in many instances, would have probably disappeared by the agency of natural elimination. The importance of a partly endogenous, partly exogenous internal rhythm of reproduction has been stressed, and it is suggested that the stable endogenous component is adaptive and has been varied in its duration from species to species according to habit and environment. Where the environment is so fruitful that the young of more than one season a year are supportable (e.g. the Victoria Nyanza in the case of cormorants), or where the environment is generally so unproductive that it is essential that reproduction should occur whenever rain falls (see below), the length of the post-nuptial regeneration period is often shortened by natural selection. The length of this period, too, is possibly influential in the timing of trans-equatorial migration.

The more prolonged exogenous component is, of course, modifiable by week-to-week events in the environment such as light (in many species), temperature (in almost all non-tropical species), danger, food and water supply, nest-site availability and behavioural reactions between the sexes. It is only when we study reproductive processes by the use of changing (and, whenever possible, unstable) environments (ideally over a succession of years) as a kind of natural laboratory that we see clearly the operation of factors such as those mentioned above.

Even in climatically stable areas the mean reproduction date of a sedentary species may vary by as much as a month from one year to another. It is believed that in most species, whether the post-regeneration cycle is initiated by light or not, the sexual cycle is 'timed' by external stimuli presented by the traditional breeding ground to the female exteroceptor organs. In all species yet investigated the males quickly achieve gametogenesis, far in advance of the female. Not until the environment presents the 'species requirement' that she innately requires will the female's cycle culminate and make reproduction possible.

The sexual cycles of some species appear to be accelerated by

rainfall, or by its effects (such as the growth of herbage or the rise of water level in lagoons) and, of course, by the increase (but not, in birds, so far as is yet known, by the decrease) of day length. The advantages of such mechanisms have been outlined. A consideration of these *various* complexes reveals clearly how completely the *biounit* – the whole animal, not merely the sexual cycle – is involved. Thus, for example, the widely unrelated finch *Quelea quelea* and duck *Anas gibberifrons*, which breed swiftly and opportunistically as soon as chance rainfall renders their unpromising environments propitious for reproduction, have correspondingly abbreviated their sexual displays, the latter in startling degree by comparison with other Anatinidae.

If we consider all the implications of, for example, innate nomadism and traditional species requirement of the female on the breeding ground, we will see that neural adaptations too, varying between closely related species and perhaps races, have been dictated by the environment. The interplay of a variety of factors has enabled *A. gibberifrons* probably to become the commonest duck in Australia and *Q. quelea*, in terms of biomass, perhaps the most successful finch in Africa. The latter is such a plague-pest that the Federation of Rhodesia and Nyasaland, the French Republic and Community, Belgium, Ghana, Guinea, Liberia, Portugal, the Union of South Africa and the United Kingdom all contribute funds to the *Commission de Coopération Technique en Afrique au Sud du Sahara* to be used for its destruction. In their millions, the highly nomadic adult finches can subsist essentially on dry seeds during the arid season and, having quickly passed through the regeneration phase, rapidly reproduce during any subsequent period of the year in which rain falls and causes the environment to present optimum amounts of nest material and, it follows, associated proteinous insect food on which the young subsist for the first five days of life.

On the other hand, many species of the same environment remain relatively unsuccessful and some, no doubt, have disappeared altogether.

The day is almost here when the study of avian breeding seasons will cease to be an endocrinological problem: it is even now essentially the business of the ecologist, ethologist and neurophysiologist.

13 M. B. Wilkins

Biological Clocks

M. B. Wilkins, 'Biological clocks', *Advancement of Science*, March 1968, pp. 273–83.

Introduction

At the present time, when the traditional and insular approach so characteristic of scientists in the various branches of biology over the last half-century is rapidly and rightly being abandoned, the study of biological clocks offers some unique opportunities for fruitful collaboration not only between biologists trained in different fields, but also between biologists, chemists and mathematicians. While the clock system itself is of very considerable interest to the pure scientist, it is becoming increasingly obvious that a greater understanding of the clock system and the part it plays in regulating the metabolism and physiological activity of an organism is of great practical importance in the medical, sociological and behavioural sciences, and in commerce.

Animals and plants at all levels of organization from man to the fungi appear to have developed the ability to measure the passage of time, and to regulate their behaviour and metabolism on a temporal basis. Biological clocks are found in both multicellular and unicellular organisms, and in man there is some evidence of several independent clock systems being present in one individual.

Space will obviously not permit a detailed treatment of all the interesting aspects of biological clocks that have emerged from recent research. The time-measuring mechanisms in living systems will therefore be discussed in general terms in an attempt to highlight progress made in recent research, and the growing realization of the importance of biological clocks in modern society.

Development of the biological-clock concept

The concept of the biological clock has arisen from three distinct lines of research. These are the studies of (1) photoperiodic phenomena, (2) navigation and orientation and (3) endogenous rhythms.

Broadly, a photoperiodic response is the reaction of an organism to changes in the relative lengths of the light and dark periods during the twenty-four-hour day. This phenomenon was discovered in plants in 1920 by the American scientists Garner and Allard who found that day length controlled the onset of flowering in a number of species of plant. Plants were exposed either to long days of sixteen hours light and eight hours darkness, or to short days of eight hours light and sixteen hours darkness. Some plants, for example *Hyocyamus niger*, remained entirely vegetative in short days but developed flowers after transfer to long days. In contrast, other species such as *Xanthium pennsylvanicum* and *Kalanchoë blossfeldiana* remained vegetative in long days and developed flowers in short days. In addition to these two classes of photoperiodic response, another group of plants appears to have complex day-length requirements for the induction of flowering. The *Chrysanthemum*, for example, requires exposure first to long days and then to short days. Many other developmental processes in plants are controlled by day length, for example tuber formation, bulbing and cambial activity.

Although plants tend to be referred to as long-day plants or short-day plants according to the conditions under which they develop flowers, recent research has shown that plants actually respond to the length of the night. Furthermore, the critical night length which determines whether or not a plant will flower does not change appreciably in plants kept at different ambient temperatures. In other words, plants showing photoperiodic responses have the ability to measure with a high degree of accuracy, over a wide range of temperatures, the length of time for which they are exposed to darkness.

A number of photoperiodic phenomena are found in animals. The development of gonads and migratory restlessness in birds has recently been shown to be controlled by the length of the day. In other vertebrates, there is evidence of photoperiodic control of metamorphosis and testicular activity in frogs, of reproduction

cycles in turtles and several species of lizard, and of ovarian cycles, oestrus and moulting in a number of mammalian species. In insects, diapause is controlled by day length.

In all photoperiodic responses the length of the night or day is measured with a remarkable degree of accuracy despite changes in the ambient temperature. An accurate measurement of time is possible only when a reliable reference system which operates at a uniform rate despite changes in temperature is available. The precision with which living organisms can detect changes in the length of the day or night has led to the conclusion that photoperiodic phenomena must be based on a reliable internal time-measuring system – the biological clock.

The classic experiments on animal navigation and orientation carried out in the early 1950s with bees and birds established that these animals use the position of the sun in direction finding. In order to operate a sun-compass the animal must be able to compensate for the relative movement of the sun across the sky during the course of the day. Such a compensating mechanism must operate at a uniform rate despite changes in environmental parameters such as temperature, otherwise large and possibly fatal errors in navigation will result. A reliable time-measuring system – the biological clock – thus seems to be an essential component of the direction-finding mechanism in animals.

That animals have a reliable clock mechanism with which they compensate for the movement of the sun across the sky has been shown by two types of experiment. In the first, animals trained to fly in a particular direction for food were removed from their natural environmental conditions and exposed to an artificial static 'sun'. The direction in which the animals flew to find food was then determined as a function of time of day. The direction taken by the animals deviated from the trained direction according to the time of day, and the angle of deviation corresponded exactly to that expected from the rate of the relative movement of the sun across the sky. In the second type of experiment, animals were again trained to travel in a particular direction to find food. They were then exposed for a week or so to artificial cycles of light and darkness which were several hours ahead or behind those of the natural environment. The animals were returned to their natural environment and the direction which they

took to find food was determined. Experiments with bees and lizards have shown that the animals took a direction which deviated from the trained direction by an amount entirely predictable on the basis of the number of hours by which their internal clock systems had been advanced or retarded by the artificial cycles of light and darkness.

It was the study of endogenous rhythms in plants that gave the first clue to the existence of the biological clock. Nearly two thousand years ago, Pliny recorded that leaflets of many plants opened during the day and closed during the night. The periodic movement of leaves attracted much attention during the eighteenth century and in 1729 the Frenchman de Mairan noticed that the rhythms of leaf movement continued even after plants had been transferred to prolonged darkness in a cellar. This was the first demonstration that the rhythmic movement of leaves was not merely the response of the plant to cyclic changes in the light intensity and temperature of the environment, but was brought about by a regulating system within the plant itself.

Since these early reports a large number of plants and animals have been found to retain a rhythmic pattern of behaviour for days or weeks after being transferred to a uniform environment. Persistent rhythms have been reported in almost all the major groups of plants and animals, and in many different physiological and chemical processes. Some of the best-investigated rhythms are those in the running activity of cockroaches, lizards and golden hamsters, in the eclosion of *Drosophila* flies from their pupae, and in the excretion of ions and water by man. In plants, rhythms of leaf movement in runner-bean seedlings, of petal movement in *Kalanchoë*, of growth rate in the shoots of young oat seedlings and of spore discharge in several species of fungi have been studied in detail. More recently, rhythms of photosynthesis and luminescence (Figure 1) in algae and marine dinoflagellates have been studied at the chemical level, as has the rhythm of carbon dioxide metabolism in the succulent plant *Bryophyllum fedtschenkoi* with which my own research has been concerned over the past twelve years. The characteristics of these persistent rhythms strongly indicate that they are controlled from within the organism by a temperature-compensated, oscillating system – a biological clock.

In this paper the physiology and chemistry of biological clocks will be discussed with particular reference to studies that have been made on persistent endogenous rhythms. Only rhythms having a period of approximately twenty-four hours will be considered. These are now termed circadian (*circa* – about, *dies* – day) rhythms.

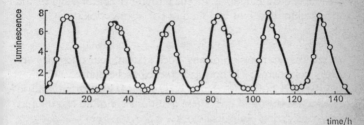

Figure 1 The persistent circadian rhythm of luminescence in a culture of *Gonyaulax polyedra* maintained in constant low-intensity light and a uniform temperature

Evidence for the endogenous nature of rhythms

There has been considerable debate over the past fifteen to twenty years on the question of whether the observed persistent rhythmicity is a true reflection of the operation of an internal oscillating system and not merely the response of the organism to some subtle periodicity in an environmental parameter of which we may even by unaware.

In 1954, Pittendrigh suggested that to establish the endogenous nature of a rhythm it should conform to the five conditions set out below. Pittendrigh's views are now widely accepted. Briefly, these conditions are: (1) that the rhythm should persist in an environment in which as many parameters as possible are held constant; (2) that the phase of the rhythm should be able to be adjusted by suitable treatment and that the rhythm should retain the new phase when persisting in a uniform environment; (3) that the rhythm should be initiated by a single stimulus; (4) that the phase of the rhythm should be delayed when the metabolism of the organism is interrupted by, for example, exposure to anaerobic conditions; and (5) that the period of the rhythm should *not* be exactly twenty-four hours. It may not be possible to determine

whether the rhythm in a particular organism conforms to all these conditions. For example, raising some plants and animals under uniform environmental conditions is technically impossible and so whether or not a rhythm is initiated by a single stimulus cannot be assessed. Almost all the rhythms that have been studied in detail conform to these conditions and are thus considered to be endogenous. Examples of such rhythms conforming to the various conditions will be found in the next section.

Physiology of the biological clock

In a critical examination of any clock system, four questions are usually considered: (1) how can the clock be started? (2) how can it be reset? (3) how reliable is it? and (4) how often, and in what way, is it wound up? These questions are of no less interest when considering biological clocks.

All biological clocks require a stimulus to start them. This need only be a single stimulus; no repetition is required to 'teach' the organism the period of oscillation. Cultures of pupae of the fruit-fly *Drosophila* which are kept in total darkness show an approximately uniform rate of eclosion. After a single exposure to light the eclosion pattern changes to a rhythmic one in which the flies emerge from their pupae only at definite times during the twenty-four-hour day. This is illustrated in Figure 2. Similarly, oat seedlings grown in continuous red light do not exhibit a rhythmic pattern of growth. On being transferred to continuous darkness, however, a rhythm in growth rate is initiated. In other organisms a single change of temperature suffices to start a circadian rhythm.

The initiation of a rhythm by a single stimulus suggests that an oscillating system with a natural frequency of about twenty-four hours is inherited from one generation to the next, but that the system does not operate until stimulated by an environmental parameter. Further support for the clock system being an inherited characteristic is the finding in the runner bean of distinct strains with significantly different natural frequencies, and the observation that while the period of rhythms in plants and animals can be changed by exposure to non-twenty-four-hour cycles of light and darkness, the natural period of oscillation reappears as soon as the organism is restored to a uniform environment. Such entrainment of rhythms indicates that while environ-

mental parameters can influence the clock system, they have no permanent effect on the system and certainly do not alter its natural frequency of oscillation.

The biological clock can be reset by a number of environmental stimuli, for example light, temperature change and exposure to low partial pressures of oxygen. The resetting of the biological clock by a particular stimulus is studied by determining the phase

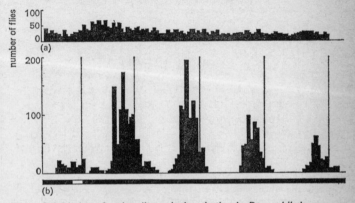

Figure 2 Initiation of a circadian eclosion rhythm in *Drosophila* by a single four-hour exposure of a dark-grown culture of pupae to light. (a) Control in continuous darkness. (b) Rhythm initiated by light stimulus. Vertical lines are twenty-four hours apart. The time of light stimulus is shown by the bar below the figure

shift induced in an observed rhythm under constant environmental conditions.

When an organism is exhibiting a rhythm in prolonged darkness a brief exposure to light will cause a predictable alteration in the phase of the rhythm. However, the magnitude of the phase shift is dependent upon a number of factors. In the case of a light stimulus these include (1) the position in the cycle at which the stimulus is given, (2) the duration of stimulation and (3) the wavelength of light employed. The importance of the position in the cycle at which a stimulus is applied is illustrated in Figure 3. Leaves of *Bryophyllum* exhibit a persistent rhythm of carbon dioxide metabolism when kept in prolonged darkness and a uniform temperature. Exposure of the leaves to light for six hours

between the peaks of carbon dioxide output induces a phase shift whereas this treatment has no effect when applied at the apex of the peak. Closely similar results have been obtained for the rhythms of leaf movement in runner-bean seedlings, and of luminescence in the armoured marine dinoflagellate *Gonyaulax*.

The duration of the stimulus determines whether or not a phase shift is induced in *Bryophyllum* leaves. Exposure to light for one

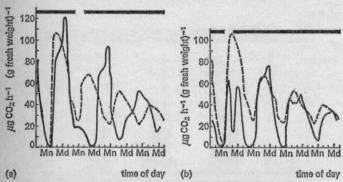

(a)   time of day   (b)   time of day

Figure 3 The effectiveness of a six-hour exposure to light in resetting the phase of the rhythm of carbon dioxide output in leaves of *Bryophyllum* otherwise kept in continuous darkness and at a constant temperature of 26 °C. Time of treatment is shown by the bar above each figure. Rhythm in untreated leaves is shown by the broken line

hour between the peaks does not shift the phase of the rhythm, but three- to six-hour stimuli are effective. The magnitude of the phase shift induced is unrelated to the length of the light stimulus; three- and six-hour light treatments induce the same shift providing they end at the same time.

In investigating a photobiological response it is important to identify the pigment or pigments responsible for capturing the light energy and hence perceiving the stimulus. This is usually achieved by determining the relative efficiency of various narrow bands of the spectrum in bringing about the response, and then comparing this action spectrum with the absorption spectra of the pigments present in the plant or animal. In higher plants, such as the runner bean and *Bryophyllum*, it has been found that the effects of light on rhythms are confined to the red end of the spec-

trum between 600 and 700 nm. In the runner bean it has been further shown that the effectiveness of an exposure to red light can be negated by immediate subsequent exposure to far-red light, that is a band between 700 and 800 nm. This type of re-

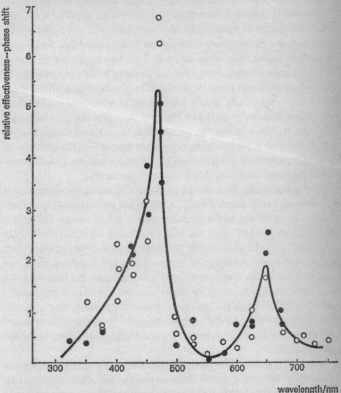

Figure 4 The action spectrum of a phase shift in the rhythm of luminescence of *Gonyaulax polyedra*

sponse is characteristic of those brought about by interconversion of the chromoprotein phytochrome which plays a vital role in controlling growth and development in plants. The one organism in which a detailed action spectrum has been determined is the photosynthetic marine protozoan *Gonyaulax polyedra*. The

spectrum, shown in Figure 4, has two peaks, a major one in the blue region and a minor one in the red region. Although at first sight the action spectrum suggests that chlorophyll may be involved in the perception of the stimulus, close inspection reveals that the peaks do not correspond exactly with those of any of the known chlorophylls, and it is therefore impossible to identify with certainty the pigment involved in photon capture. In the fungi, white light is active in initiating the rhythm and in changing the phase. As far as the author is aware, red light has no effect on fungi. The inference is, therefore, that the activity of light in modifying the fungal rhythms must be confined to the blue end of the spectrum. This obviously needs experimental verification. The fact that quite different regions of the spectrum are active in controlling the rhythms in higher plants, *Gonyaulax* and fungi, clearly indicates that different pigments are responsible for perceiving the light stimulus. Unfortunately, in no organism has the pigment yet been identified with absolute certainty.

Phase shifts can be induced in plant and animal rhythms by a number of other stimuli such as exposure to high and low temperature, or to anaerobic conditions for a few hours. As in the case of exposure to light, the effectiveness of all these treatments depends upon the position in the cycle at which they are applied. In *Bryophyllum*, raising the temperature from either 16 °C or 26 °C to 36 °C for three or six hours induces a phase shift which is identical to that induced by exposure to light. The treatment is effective in shifting the phase in exactly the same parts of the cycle as exposure to light and the magnitude of the shift is determined by the time the treatment ends. These points are illustrated in Figure 5.

Thus, although the biological oscillating system operates independently of the cyclical fluctuations in environmental parameters, the parameters do set the phase of the system and hence act as time clues or correction stimuli for the biological clock.

The reliability of the biological clock is of paramount importance and, as with mechanical clocks, the most likely source of error is variation in ambient temperature. This might be especially expected in a biological clock which must be based on chemical reactions which normally double or triple their rate with a rise in temperature of 10 °C.

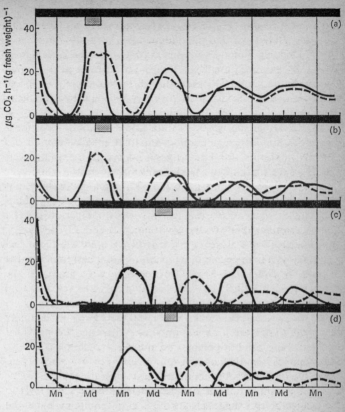

Figure 5 The effect of the raising of the temperature to 36 °C for three or six hours on the phase of the rhythm of carbon dioxide metabolism in leaves of *Bryophyllum* otherwise kept in darkness and at 16 °C. Times and duration of treatments are shown by the hatched area in each figure. Rhythms in untreated leaves are shown by the broken lines

Probably the most remarkable feature of biological clocks is that they obviously incorporate some kind of temperature compensating mechanism which reduces, and in some cases eliminates, variation in the period of oscillation with changes in temperature. In runner-bean and oat seedlings no significant change

in the period of oscillation could be detected over the temperature range between 15 °C and 30 °C. On the other hand, the period of the rhythm of carbon dioxide metabolism in *Bryophyllum* is 22·4 hours at 26 °C and 23·8 hours at 16 °C. This change is significant but extremely small. In *Gonyaulax* there is also a small change in the period of the rhythm of luminescence, but in the opposite direction to that found in *Bryophyllum*, the period being 25·5 hours at 32 °C and 22·5 hours at 16 °C.

The facts that the period of oscillation is slightly dependent upon the ambient temperature, and that it is different from twenty-four hours, are the strongest evidence for the endogenous nature of the rhythmicity. Under uniform environmental conditions, runner beans with a period of approximately twenty-eight hours are daily falling behind the environmental cycles by about four hours, whereas *Bryophyllum* leaves with a period of 22·4 hours are gaining daily on the environmental cycles by 1·6 hours.

An interesting feature in *Bryophyllum* is that the biological clock shows a high degree of temperature compensation over the approximate range 12–31 °C. Outside this range, however, the clock stops altogether, but will begin again when restored to a temperature between 12 and 31 °C.

As with mechanical clocks, biological clocks require a supply of energy to operate. In living organisms this energy is provided by metabolism. In both plants and animals, the biological clocks are stopped when metabolism is interrupted by exposing organisms to either anaerobic conditions or temperatures near 0 °C.

Closer examination of the dependence of the clock system upon metabolism suggests that the energy requirement may be different in different parts of the cycle. In *Bryophyllum* and bean seedlings, for example, it is possible to deprive the plants of oxygen for four to six hours in some parts of the cycle without delaying the phase of the rhythms, whereas at other parts of the cycle a phase delay is induced. These results, shown in Figure 6, suggest that over certain parts of the cycle of oscillation the clock system operates independently of metabolic energy. Generally similar results from experiments in which exposure to low temperature was used led to the suggestion that the oscillating system of the biological clock was of the relaxation-oscillator type. However, no such simple explanation is acceptable because it has been

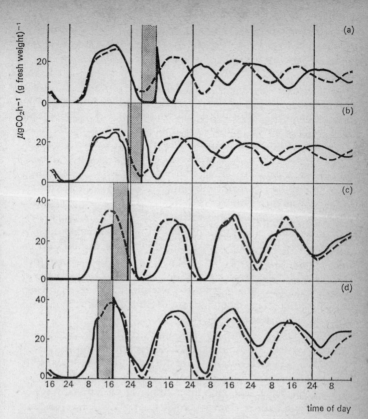

Figure 6 The effect on the phase of the rhythm of carbon dioxide metabolism of exposing leaves of *Bryophyllum* to anaerobic conditions for six hours at different times of the cycle. Position of treatment is indicated by the shaded zone. Rhythms in untreated leaves are shown by broken lines

found that the parts of the cycle over which exposure to low temperature and anaerobic conditions are without effect on the clock system are not identical. Both low temperature and lack of oxygen might have been expected to exert a similar influence since both reduce to a low level the rate of metabolism of the tissues.

The clock system is obviously dependent upon metabolism for

energy supply, but the details of whether the supply of energy must be continuous or only over certain parts of the cycle await further investigation.

Location of the biological clock and its chemical mechanism

In unicellular organisms there is little doubt that each cell has a circadian system controlling the rhythms manifest in its physiological and chemical activity. In multicellular organisms, however, the question arises as to whether the circadian system is confined to specialized cells, as seems to be the case in some animals, for example the cockroach, or whether the activity of each cell in the organism has its own controlling circadian system. Evidence obtained from the rhythm of carbon dioxide metabolism in leaves of *Bryophyllum* suggests that each cell has its own circadian system. Small cubes of leaf mesophyll from which the epidermis has been removed still exhibit circadian rhythms of carbon dioxide metabolism regardless of the region of the leaf from which they are taken. Furthermore, tissue cultures of leaf mesophyll cells of *Bryophyllum* also show circadian rhythmicity in carbon dioxide metabolism, so in this plant at least, rhythmicity does not depend upon the organization of the leaf.

In other higher plants it is not clear whether the controlling circadian system is localized. In runner-bean seedlings, for example, the periodic changes in the turgidity of the cortical cells of the pulvinus that give rise to the rhythm of leaf movement appear to be dependent upon the presence of the lamina. It is not known whether the rhythm in turgidity of the cortical cells is due to a rhythm in the synthesis of a growth hormone in the lamina, or in the transport of the hormone from the lamina to the pulvinus.

Locating the clock system within the cell is of fundamental importance since it demands the elucidation of the clock's chemical mechanism.

Three organisms have been investigated with a view to ascertaining the intracellular location of the circadian oscillating system. The umbrella-shaped, giant unicellular alga *Acetabularia* has certain advantages as an experimental material in this type of investigation, since the nucleus is conveniently located at the base of the stalk and can therefore be removed easily. It has been found

that a rhythm in photosynthetic capacity continues in enucleate and normal cells in a similar manner. Furthermore, the phase of the rhythm is easily shifted by light, regardless of whether the nucleus is present in the cells or not. This evidence suggests that the basic clock system is not located in the nucleus, and that the nucleus is not essential for mediating in the light-induced phase shifts. To assume that the nucleus has no role in the circadian system would, however, be unjustified. When cells of *Acetabularia* are grown in cycles of light and darkness differing in phase by twelve hours, the phases of the rhythm of photosynthetic capacity of the two stocks differ by twelve hours. If nuclei from one stock are now transplanted into cells of the other stock and the stocks then transferred to a uniform environment, the phase of the rhythm in those cells that have been subjected to nuclear transplantation corresponds to that of the light–dark cycle to which the nucleus has been subjected. The circadian oscillating system thus appears to be located in the cytoplasm or cytoplasmic organelles, and can operate entirely independently of the nucleus. When the nucleus is present, however, it apparently plays a dominant role in controlling the phase of the cytoplasmic oscillating systems.

In enucleate cells of *Acetabularia*, the persistence of the rhythm, and the sensitivity of the phase of the rhythm to light stimuli, do not establish that nucleic acid metabolism is not involved in the biological-clock system since considerable amounts of DNA have been found in the chloroplasts and synthesis of RNA has been demonstrated in cells from which the nucleus has been removed. However, the phase and period of the rhythm in photosynthetic capacity are unaffected by the high concentrations of actinomycin D, pyromycin and chloramphenicol, which markedly reduce RNA and protein synthesis. These results point to the circadian rhythmicity being independent of nucleic acid and protein synthesis. Only a small fraction of the total nucleic acid and protein synthesis might, however, be involved in the circadian oscillation system and a considerable fraction of RNA metabolism is unaffected by actinomycin D, since total inhibition of RNA synthesis is never achieved with this inhibitor. So even these experiments do not provide unequivocal evidence which can resolve the problem of the extent to which nucleic acid metabolism and

protein synthesis are involved in the circadian oscillating system in *Acetabularia*.

It would be extremely valuable to know more of the chemical changes which accompany the rhythm of photosynthetic capacity in cells of *Acetabularia*. For example, information on whether or not the amounts and specific activities of the enzymes responsible for photosynthetic carbon dioxide fixation vary rhythmically would indicate whether rhythmic enzyme synthesis is involved. If periodic enzyme synthesis were involved then obviously much interest must centre on nucleic acid metabolism being involved in the basic oscillating system.

In two other organisms attempts have been made to analyse the chemical changes occurring in cells exhibiting circadian oscillations. In *Gonyaulax*, the rhythms of both luminescence and photosynthesis appear to be due to periodic changes in the level of enzymes involved in these reactions. The specific activity and total activity of luciferase has been shown to vary diurnally, and assays of ribulose-diphosphate carboxylase from homogenates of *Gonyaulax* cells showed a distinct cycle of activity. These results suggest that there is a periodic synthesis of luciferase and ribulose-diphosphate carboxylase in *Gonyaulax*, and one explanation for this periodic synthesis of enzyme is that there is a periodic synthesis of messenger RNAs coding for these two enzymes. If this were the case, inhibitors of RNA synthesis, such as actinomycin D, and of protein synthesis, such as puromycin and chloramphenicol, ought to inhibit the rhythm and induce phase shifts. The effects of these substances on the rhythm in *Gonyaulax* have been investigated and it has been found that actinomycin D abolishes the rhythm in luminescence. The inhibition is not, however, immediate; one peak of luminescence occurs after the beginning of the treatment. This finding suggests that if messenger-RNA synthesis in *Gonyaulax* is totally blocked by actinomycin D, synthesis of the RNA responsible for a particular peak of luminescence takes place some twenty-four hours earlier. In other words, RNA synthesis occurring at a particular time commits the cells to a programme of luminescence for the next twenty-four hours. The action of actinomycin D supports the idea that the peaks of luminescence are due to rhythmic *de novo* synthesis of

luciferase. However, the effects of inhibitors of protein synthesis only partially support this idea. Puromycin caused an almost immediate abolition of the rhythm, a finding consistent with the general scheme outlined above, but chloramphenicol had the effect of enhancing the amplitude of the rhythm which otherwise persisted normally! Substances which inhibit DNA synthesis ought to have little or no effect on the rhythm, if the rhythmicity is based on periodic RNA synthesis. Mitomycin, which inhibits DNA synthesis, had no effect on the rhythm of luminescence until about three days after application began, but then the rhythm was abolished. If DNA turnover occurs at a low rate, the delayed inhibition of the rhythm may be due to the lack of newly synthesized DNA to replace that broken down.

Investigation of the rhythm of carbon dioxide emission by leaves of *Bryophyllum* has begun at the chemical level. The rhythm has been shown to be due to the periodic activity of a carbon dioxide fixation system in the leaves. Several enzymes have the capability of bringing about carbon dioxide fixation in these leaves in total darkness but the enzyme most likely to be principally involved is phosphoenolpyruvic carboxylase. Studies have been made of the causes underlying the rhythmic activity of the system responsible for carbon dioxide fixation in the leaf. There are large amounts of phosphoenolpyruvic carboxylase present in the leaf, even in those phases of the cycle in which the carbon dioxide fixation activity of the leaf is zero. In contrast to the findings on *Gonyaulax*, therefore, the rhythm of carbon dioxide metabolism in *Bryophyllum* does not appear to be due to rhythmic variation in the amount of enzyme in the leaves. Some other mechanism must clearly be involved – cycles in the availability of substrates for the carboxylation reaction, or in the concentration of an inhibitor of the fixation reaction, are the two most obvious possibilities. When carbon dioxide concentration is non-limiting, infiltration of leaves with phosphoenolpyruvic acid at times when carbon dioxide fixation was zero caused no increase in fixation. It seems unlikely, therefore, that substrate concentration is limiting when fixation by the leaves is extremely low. Possible periodic variation in the availability of coenzymes has not yet been investigated. However, the rate of oxygen uptake by

Bryophyllum leaves in darkness shows no rhythmic variation. The periodic appearance of an inhibitor of phosphoenolpyruvic carboxylase in the tissues has been examined. At least two inhibiting substances are present in the leaves, but they are present at all phases of the cycle of carbon dioxide fixation. One of these substances appears to be citric acid.

Interpretation of these results is rather difficult at this stage of the investigation but a tentative conclusion is that at least one of the inhibitors arises as an end-product of the carbon dioxide fixation reaction. Further fixation might then be prevented until the concentration of inhibitor at the site of fixation is reduced by removal of the inhibitor to a new intracellular location.

The present primitive state of our knowledge of the chemistry of the rhythm in *Bryophyllum* obviously permits only a speculative assessment of the mechanism involved. The absence of periodic fluctuation in extractable enzyme activity does, however, point to the mechanism being somewhat different from that involved in *Gonyaulax*. At present there appears to be no evidence to suggest that nucleic acid metabolism is intimately involved in the basic oscillating system in *Bryophyllum* leaves.

The practical importance of biological clocks

Considerable survival value stems from the operation of the biological clock in controlling and correlating physiological activities in living organisms. For example, emergence of *Drosophila* flies from their pupae occurs just before dawn when the humidity is highest and they have the greatest chance of avoiding desiccation. The time of emergence is determined by the biological clock being set by the previous dusk. Similarly, synchronization within the biotic environment is of importance, especially between sexual partners.

Biological clocks are of very considerable practical importance in the medical and sociological sciences, and in commerce. The level of tolerance of animals to drugs and ionizing radiation changes according to the time of day. In at least one animal, the cockroach, the importance of the organization of the physiological functions of the body on a twenty-four-hour cycle has been clearly demonstrated. In this animal the clock system is localized in the suboesophageal ganglion. This organ can be trans-

ferred from one insect to another and the rhythmic activity of the receiving insect assumes the phase of the rhythm in the donor insect. In other words, biological clocks can be transferred from one insect to another. Up to four ganglia have been implanted in one insect without harmful effects, providing the individual clocks were in phase, that is, essentially, all telling the same time. In marked contrast to this, if a ganglion is transferred to a cockroach whose own clock system is twelve hours ahead or behind that in the implanted ganglion, a pathological condition arises in the form of massive and fatal midgut tumours.

In man there are circadian rhythms in a number of physiological functions, for example, body temperature, ion and water secretion and pulse rate. Observations on submariners, cave dwellers and subjects confined to uniform environmental conditions have shown these rhythms to be quite independent of the cyclical changes in light intensity and temperature in the normal environment. The phases of the rhythms in the different physiological functions of the body normally bear a definite relationship to each other. The relationship can, however, be upset. Experiments in Spitzbergen during the summer with subjects living on an artificial twenty-seven-hour day revealed that the water-secretion rhythm soon adapted to a twenty-seven-hour cycle, whereas the potassium-secretion rhythm remained on a twenty-four-hour cycle. The relative times of occurrence of the peaks were therefore changing on successive days.

In the modern world there are a number of situations in which the normal synchronization between the rhythms of the body and the environment can be upset. The most important of these is the change in the daily pattern of activity and sleep which occurs after a change of shift in a factory, hospital or police station, and after a long journey in a fast jet aeroplane in an easterly or westerly direction. In both instances, the phase of the body rhythms have to adjust to the new cycle of activity and sleep and this does not occur at once. In addition, some body rhythms may adjust to the new phase of the activity rhythm more rapidly than others and so the individual is contending not only with the body rhythms being out of phase with his new pattern of activity and sleep, but also the different body rhythms being out of phase with each other. Individuals vary a good deal in the rate at which their

body rhythms synchronize with a shifted activity rhythm; however, most seem to achieve synchronization within one to five days.

According to a Ministry of Labour census the number of people working on a rotating shift system doubled between 1954 and 1964, and the number of people engaging in jet travel has increased enormously during that period. It is of importance, therefore, to ascertain what effects these changes in the daily pattern of sleep and activity, and hence desynchronization of body rhythms, have on the efficiency and health of the individual.

Observations on both shift workers and travellers have shown that after a shift in the regimen of activity and sleep, a high degree of fatigue, a marked increase in decision time, and a marked decrease in proficiency occurs. The subjects recover their normal proficiency and decision times after being in the new regimen for one to five days, but the rate of recovery varies between individuals. The rate of recovery appears, however, to be correlated with the rate at which the physiological rhythms of the body synchronize with the new pattern of activity.

Obviously this temporary decrease in proficiency is of importance to the manufacturing industries where it may result in a rise in the proportion of unacceptable items from a production line for the few days after a change-over in shift. Similarly, at the administrative level, major policy decisions and negotiations should not be undertaken by executives or diplomats after a long east or west flight until they have been in the new location for a few days to recover their normal levels of proficiency.

What effect, if any, repeated shifting in the rhythmic pattern of activity and sleep have on the health of man is not entirely clear at the present time, but there are reports of a higher incidence of body discomfort, constipation and indigestion in workers on rapidly rotating shifts. One study suggested a higher incidence of stomach ulceration, rheumatoid arthritis and upper respiratory infection in shift workers. However, owing to the different rates at which rhythms adapt to a new pattern of activity and sleep, some people are undoubtedly better suited to shift work and rapid east–west travel than others. This aspect of industrial medicine illustrates that quite apart from their intrinsic interest, the oscillating systems which we refer to as biological clocks have an important impact on modern society.

14 D. Noble

The Initiation of the Heartbeat

D. Noble, 'The initiation of the heartbeat', *Advancement of Science*, December 1966, pp. 412–18.

Introduction

The heart of a healthy person beats once every second for an average of seventy years. Various factors can influence this rate. The heart beats faster, for example, during exercise and when someone is excited. It is therefore under some form of nervous or hormonal control. However, the origin of the beat lies in the heart itself and the heart will continue beating regularly even when isolated from the rest of the body and, therefore, from all the usual forms of control.

The nature of this autorhythmicity has interested physiologists for a long time. Leonardo da Vinci (1452–1519) observed that the heart 'moves by itself' and he probably also realized that it drives blood into the arteries, although he did not know that its activity causes the blood to circulate. Even after the discovery of the circulation by Harvey (1628) physiologists could only explain why (i.e. for what purpose) the heart beats rhythmically; the mechanism of initiation of the rhythm was completely unknown and, for many biologists, the inherent activity of the heart was a major example of the working of the vital forces thought to be characteristic of living organisms. In fact, the heart was thought to be the source of the vital 'spirits', a view the persistence of which seems to be mainly attributable to the erroneous belief that the heart is the source of heat in the organism. Although Harvey himself questioned this view and clearly stated that the heart appears to be the source of heat only because the blood it receives is already warm, the controversy continued right into the eighteenth and nineteenth centuries and was only fully resolved when the activities of living systems were systematically subjected to analysis in physicochemical terms. One of the key figures in

this development was Bernard whose lifetime (1813–78) saw a complete revolution in physiology and, in particular, the resolution of the mechanist–vitalist controversy, largely as a result of Bernard's own work. The view that the heart's activity is the manifestation of vital forces has therefore been generally abandoned amongst physiologists for a long time, although it is only relatively recently that we have been able to investigate the mechanism of cardiac rhythmicity in a way which may provide an adequate physicochemical explanation. In this paper I shall try to show how far we have got towards such an explanation.

The electrical beat of the heart

The heartbeat is, of course, a mechanical event. It consists of the pumping action of the heart muscle which forces the blood round the circulation. But it is also an electrical process. Each time the muscle contracts electric currents flow through it. Since the body

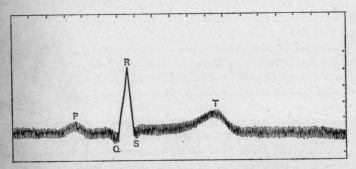

Figure 1 Electrical potentials recorded between the surfaces of the two arms using a string galvanometer. Horizontal scale: time in $\frac{1}{20}$ and $\frac{1}{10}$ s; vertical scale: potential in $\frac{1}{2}$ and 1 mV.

in which the heart lies is a salt solution which can conduct electricity, some of these electric currents can be detected in the body a long distance away from the heart itself. They can in fact be detected at the surface of the body. However, the currents at this distance are very weak and it was for the purpose of recording these currents that Einthoven developed the string galvanometer at the beginning of this century. One of his original records is shown in Figure 1. Each time the heart beats, the potential between

two points on the body surface (in this case between two arms) varies in a characteristic manner. A small slow wave (designated P by Einthoven) is followed after a fast wave (R) by another slow wave (T). These waves can in fact be associated with events occurring in the heart itself (P corresponds to auricular contraction; R and T to ventricular contraction) and, since the method is basically simple and requires no operative procedures, it is used as a clinical test to check whether the heart's activity is normal.

Unfortunately, from the physiologist's point of view, this method of recording the electrical activity of the heart is too indirect and the result which it gives is too complicated. As is frequently the case in scientific research, in order to obtain a result which is simple enough to analyse it is necessary to use more elaborate methods. In this case, we must record directly from the heart itself and, preferably, from very small parts of it.

The electrical activity of single cells

Like all organs in the body, the heart muscle is formed by a very large number of tiny units called cells. Cells are so called because they are compartments of living material separated from their environment by a thin container – the cell membrane. In electrically excitable cells, such as nerves and muscles, the cell membrane is not only the cell boundary, it is also the part of the cell which is mainly responsible for its electrical activity. Thus, if a muscle cell membrane is removed by careful dissection, the cell no longer shows electrical activity, although it is still capable of moving if appropriate chemicals are applied. It is also possible to nearly completely remove the interior contents of a nerve cell without impairing the ability of the cell membrane to show electrical activity provided that it is surrounded by appropriate salt solutions.

In order therefore to record the electrical events of interest to the physiologist, we need to record the electrical activity of the membrane of a single cell and the most direct way of doing this is to record the electrical potential difference between one side of the membrane and the other. But this requires that we should place one of our recording leads inside a single cell, and this creates a severe technical problem. Even the largest heart cells are only

$\frac{1}{20}$ mm in diameter. The recording lead must be much smaller still so that it can be inserted inside a cell without causing appreciable damage. Physiologists have achieved this by using a miniature version of an extremely common piece of scientific apparatus: the glass pipette. If a glass tube about 2 mm in diameter is melted at one point and drawn out sufficiently quickly it is possible to obtain a pipette whose tip is less than $\frac{1}{10000}$ mm in diameter. Using appropriate manipulating apparatus this pipette can be inserted into a cell without causing appreciable damage. Moreover, if the

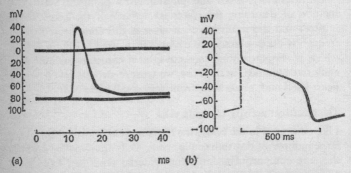

(a)

ms (b)

Figure 2 (a) Membrane potentials during activity produced by stimulating a frog skeletal muscle fibre. Note that action potential is short (about ten milliseconds). The action potentials in nerve cells are of similar shape but are even shorter. (b) Membrane potentials during spontaneous activity in a Purkinje fibre from a sheep's heart. Note that action potential is extremely long (the time scale is much compressed compared to that in (a)) and that there is no resting potential

pipette is previously filled with a salt solution, it will serve as an electrical contact with the interior of the cell. If the potential of the pipette is then recorded with respect to an electrode lying in the external fluid, we obtain a measure of the electrical potential across the cell membrane. This potential is usually called the *membrane potential*.

The membrane potentials recorded in this way in heart muscle cells are rather more complicated than those recorded in cells which do not show autorhythmicity and it will be more convenient to briefly describe the membrane potentials recorded in nerve and skeletal muscle cells first, since these cells are usually

quiescent unless stimulated. This means that we can study their potentials in the resting state. It is found that in this state the interior of the cell is at a negative potential with respect to the extracellular fluid. This potential is called the *resting potential* and is maintained so long as the cell is not excited. When the cell is excited by passing an electric current through it whose strength is sufficient to deflect the membrane potential beyond a certain critical value (the *threshold potential*), the interior of the cell rapidly becomes positive for a brief period of time, usually 1–10 ms. This transient reversal of the membrane potential is called an *action potential* (see Figure 2). In nerve cells the action potential is the signal which is transmitted from one end to the other; in muscle cells it is the event which triggers contraction.

Cardiac membrane potentials

The characteristics of the membrane potentials recorded from single heart cells depend partly on the region of the heart from which they are obtained. The largest cells lie in the fast-conducting system (the Purkinje fibres) and, since it is easier to record from these cells, they have been widely used in electrophysiological studies. They have the additional merits that they are easy to dissect out of the heart and they show most of the electrical characteristics of heart cells, including autorhythmicity. When membrane potentials are recorded in these cells we obtain a result which is striking in that it differs from that obtained in nerve and skeletal muscle cells in two important respects. First, there is usually no resting potential. At some times, the interior of the cells is negative by a potential which is similar to the resting potential in other excitable cells, but this potential is not maintained. It spontaneously changes in a positive direction until the threshold potential is reached. An action potential then occurs. The second major difference is that the action potential lasts very much longer than in nerve and skeletal muscle cells (see Figure 2). In fact it is usually several hundred milliseconds in duration. At its termination, when the internal negative potential has been restored, the spontaneous change towards the threshold recommences. This cycle of events continues indefinitely if the cell is bathed in a salt solution similar to that in which it is usually bathed in the body. Autorhythmicity is therefore a characteristic

of electrical activity at the cellular level. If we can give an account of the mechanism of these potential changes, we shall have advanced a long way towards explaining the autorhythmicity of the whole heart. It will still be necessary of course to show that action potentials arising spontaneously in single cells can spread through the whole heart and that the action potential is responsible for initiating the mechanical beat. A considerable amount of work has been done on these two aspects of the problem recently but I shall confine myself in this paper to the problem of explaining the spontaneous changes in membrane potential at the cellular level.

Membrane currents underlying membrane potentials

How are membrane potentials generated? In all cells, whether they are electrically excitable or not, the composition of the intracellular fluid differs markedly from that of the extracellular fluid. Thus, in most animal cells, the concentration of potassium (K^+) salts inside the cells is about fifty times greater than the outside concentration, whereas sodium (Na^+) salts are very much more concentrated outside. These concentration differences are created and maintained by a membrane mechanism (known as the 'sodium pump') which uses energy provided by the cell's metabolism to extrude Na^+ in exchange for K^+. This results in strong ionic-concentration gradients existing across the cell membrane. If the membrane is permeable to these ions they will tend to move across it (Na^+ inwards, K^+ outwards) and, since they are positively charged, they will carry a positive current flowing in the same direction as the ions.

Let us first suppose that the membrane is permeable to K^+ but not to Na^+. Then the diffusion of K^+ ions will create an *outward* positive current and so leave the cell with a net negative potential [. . .] which turns out to be roughly equal to (in fact slightly more negative than) the resting potential in quiescent cells and the maximum negative potential in cells showing autorhythmicity.

Now let us make the alternative assumption that the membrane is permeable to Na^+ and impermeable to K^+. The diffusion of Na^+ ions will create an *inward* positive current and will therefore give the inside of the cell a net positive potential [. . .] nearly equal

to (in fact slightly more positive than) the maximum positive potential occurring during the action potential.

During activity, therefore, the membrane potential swings from a negative potential [. . .] towards a positive potential [. . .], which suggests that the membrane is changing from a state in which it is mainly K^+ permeable to a state in which it is mainly Na^+ permeable. Mainly as a result of work done in the last twenty years by Hodgkin, Huxley, Katz, Keynes and their colleagues, the evidence in favour of this view is now very convincing.

The general mechanism of a potential change may therefore be represented as follows:

Membrane permeability change $\rightarrow$ flow of ionic current $\rightarrow$ potential change

What causes the change in membrane permeability? This is one of the key questions in electrophysiology. The peculiar characteristic of electrically excitable cells is that the membrane potential itself controls the membrane permeability. This property is of major importance in the mechanism of the action potential and was first investigated quantitatively in nerve cells by Hodgkin, Huxley and Katz in 1952. They found that when a sudden positive change in membrane potential is imposed on the cell (this may be done with appropriate electronic techniques) the membrane undergoes a rapid transient increase in Na^+ permeability. The result of changing the membrane potential beyond the threshold potential is therefore to initiate a sequence of permeability changes which first tends to move the membrane potential [first in one direction and then the other]. Hodgkin and Huxley showed that the dependence of the Na^+ and K^+ permeabilities on membrane potential could be described in terms of a set of three first-order differential equations (which probably correspond to voltage-driven reactions in the membrane) and that the solutions to these equations could fairly accurately reproduce the potential changes occurring during the action potential.

Membrane currents in heart cells

Hodgkin, Huxley and Katz's experiments were performed on nerve cells about 0·5 mm thick. It is only more recently that the

membrane currents have been determined in much smaller cells and, at present, only the largest heart cells (the Purkinje fibres) have been analysed quantitatively. As we might expect from a comparison between the potential changes in nerve and heart cells (see Figure 2), it is found that the membrane currents also show similarities and differences.

In both kinds of cell, the first effect of a positive potential change is a large transient inward flow of sodium ions. However, whereas in nerve fibres this is quickly followed by an almost equally large outward flow of K^+ ions, in Purkinje fibres the outward current is very much smaller and develops extremely slowly. This is shown in Figures 3 and 4. In Figure 3 the membrane currents in response to various magnitudes of positive potential change are shown and it can be seen that potential changes which exceed threshold (about -65 mV) produce a large 'spike' of inward (downward) current. This is then followed by a very slowly developing outward current which is probably carried by K^+ ions. The equations used by Hodgkin and Huxley to describe the ionic currents in nerve fibres predict that the speed and magnitude of the currents should both increase with the strength of the positive potential change. Although they are very large quantitative differences, this is also true of the outward current in Purkinje fibres. This is shown in Figure 4 in which the time course of onset of the slow outward current is plotted at different magnitudes of potential change.

Another important difference between the membrane currents in nerve and heart cells is that the slow increase in outward K^+ current in Purkinje fibres in preceded by a large *decrease* in K^+ permeability. This effect is not evident in the experimental records shown in Figure 3 since these show the total membrane currents and, in these circumstances, it is difficult to distinguish a decrease in K^+ permeability from an increase in Na^+ permeability. However, an initial decrease in K^+ permeability has been established on the basis of other experiments. These differences in the effect of membrane potential on the K^+ permeability are important in accounting for the extremely long duration of the cardiac action potential. The membrane potential initially remains [positive] because the K^+ permeability is reduced and returns [to negative] slowly because the increase in K^+ permeability is so slow.

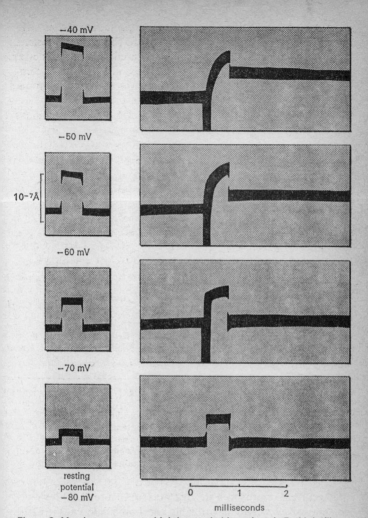

Figure 3 Membrane currents (right) recorded in a sheep's Purkinje fibre during imposed changes in membrane potential (left). Potentials which are more positive than −65 mV produce a very transient 'spike' of inward (sodium) current followed by a very slowly developed outward (probably potassium) current which decays even more slowly when the normal negative potential (−80 mV) is restored. Note that the transient inward currents are normally too fast to be photographed on slow time scales. These records have therefore been retouched to show inward currents more clearly

While the onset of the outward current in response to a positive potential change is slow compared to that in nerve fibres, its decay when the full internal negative potentials is restored is even slower and it may require one or two seconds before the mem-

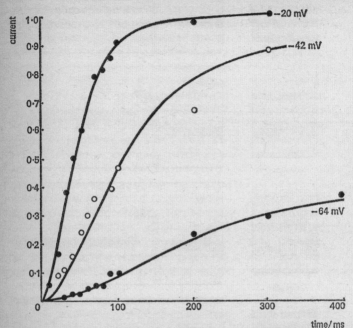

Figure 4 Time course of development of slow outward current in a sheep's Purkinje fibre in response to imposed potentials. The results were obtained by applying potentials for various periods of time and measuring the peak of the outward current following restoration of the potential to about −85 mV. The curves are calculated from equations similar to those used to describe the outward current in nerve cell

brane current returns to a steady state (see Figure 3). This slowly declining K^+ current is probably the immediate cause of the spontaneous positive potential change responsible for autorhythmicity. As the K^+ permeability declines so the potential moves [from one direction to the other] and, when the potential reaches the threshold potential, the large increase in Na^+ permeability is triggered and another action potential occurs.

Computed action potentials

If this explanation of the mechanism of potential changes in cardiac cells is correct, it should be possible to reproduce the potential changes using equations describing the membrane currents. In 1962, I described a set of differential equations similar to those used by Hodgkin and Huxley, but incorporating the

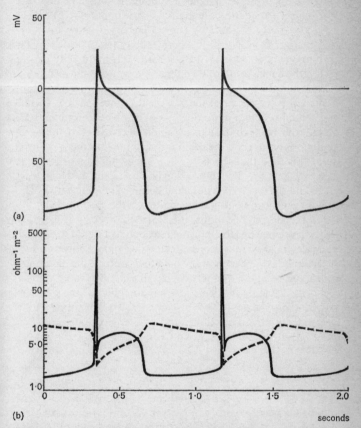

(a)

(b)

Figure 5 (a) Potential charges resembling those recorded in Purkinje fibres using a set of differential equations based on those of Hodgkin and Huxley but incorporating the major differences between nerve and heart cells. (b) Corresponding comparted conductance changes

major observed differences, i.e. the initial fall in K^+ permeability in response to a positive potential change and the greatly reduced magnitude and speed of onset of the outward K^+ current. Although subsequent experimental work has shown that these equations require quantitative modifications, they do nevertheless illustrate the theory I have described. One of the solutions to the equations, obtained on a digital computer, is shown in Figure 5. It can be seen that the potential changes (upper curve) can be reproduced fairly accurately. The lower curves show the computed changes in Na^+ and K^+ permeability (as indicated by the membrane conductance to these ions). Autorhythmicity can be seen to arise as the result of an interaction between the membrane potential and the membrane permeabilities. When the membrane potential is [very negative] the K^+ conductance declines so that the potential becomes more positive. When a certain potential is reached, the Na^+ conductance rapidly increases, an inflow of sodium ions occurs and the cell interior rapidly becomes positive. It is at this time that the fall in K^+ conductance is thought to occur. As the K^+ conductance slowly rises again, K^+ ions leave the cell and the membrane potential becomes more negative. This cycle of events can continue indefinitely.

With certain modifications, therefore, the theory of electrical activity developed by Hodgkin and Huxley may be used to account for the electrical events in cardiac cells. The major problem which now arises is how the membrane potential controls the membrane permeability. The theory suggests that this involves some voltage-dependent reactions in the cell membrane but, as yet, we know too little about the molecular structure of the cell membrane to be able to identify these reactions.

15 W. A. Van Bergeijk

Anticipatory Feeding Behaviour in the Bullfrog

W. A. Van Bergeijk, 'Anticipatory feeding behaviour in the bullfrog'
Animal Behaviour, vol. 15, 1967, pp. 231-8.

The history of conditioning of anurans is, by and large, a tale of frustration. In spite of the frog's prominence as an experimental animal in various fields of biology, the literature records an inordinately small number of papers that purport to demonstrate learning in these animals. Another small number of publications describe attempts at conditioning which failed, or at least largely so. Among the successful ones were Yerkes working with *Rana*, Goldstein, Spies and Sepinwall, also with *Rana*, Martof with *Bufo* and Bajandurow and Pegel with *Rana*; the latter's evidence, however, was called into question by Capranica. Among recent unsuccessful or partially successful attempts are those of Kleerekoper and Sibabin, McGill and Munn all working with *Rana*, the latter with tadpoles. No clear pattern emerges that discriminates successful experiments against failures; although the last three studies, for instance, all used avoidance conditioning to electric shock with little or no success. Martof obtained consistently positive results with this method.

In the spring of 1965 an opportunity arose to study a conditioned response in the colony of bullfrogs (*Rana catesbeiana*) in our laboratory. These frogs were kept for physiological experiments, and were maintained on a regular once-a-day feeding schedule. We had noticed that the animals would anticipate feeding time by gathering around the feeding trough several hours in advance; this spontaneously established behaviour appeared to present a unique chance to find out whether frogs can be conditioned.

Method

The animals on which this study was done were obtained some time in late summer 1964. The observations were carried out in the spring of 1965 with eighteen animals of 7–10 cm body length. The frogs were kept in a terrarium [Figure 1] about 1×3 m, of which one end formed a pond (1×0.6 m), and the remainder was a gently sloping beach of ordinary builders' sand. Rocks, plastic waterlilies and plastic greenery embellished the terrarium, which was covered by a coarse-mesh screen canopy, about 0.5 m high, with four access ports on one side. The water in the pond was continually changed by a slow dribble of tap water; a stand-pipe drain at the lowest end of the pond removed overflow. On either side of the beach was a stainless-steel feeding trough which rested on the sand. Figure 1 shows the terrarium in plan and rear elevation with a few features added for the experiment. The frogs were fed once every working day with a handful of live mealworms in one of the feeding troughs.

A 35 mm movie camera was rigged above the terrarium, so that it surveyed the area outlined in Figure 1, or about half of the beach. The camera was fitted with a single-frame motor, driven from a timing device that took one frame every fifteen minutes, day and night. Synchronized from the same timer was a light (ordinary 100 W incandescent) which would turn on one second before the camera clicked, and off again two seconds later.

Feeding had always been done at 4 p.m. in the trough closest to the access ports; in order to delineate the area of congregation in such a way that the site of feeding could be changed to the other trough without incurring uncertainty about the response, a 0.2 m-high barrier was erected in the middle of the beach (Figure 1), thus creating Area I and Area II. This is, in effect, a Y-maze.

The developed negative was projected frame-by-frame, and the frogs visible in Areas I and II were counted. The numbers so obtained were tabulated and punched on IBM cards together with relevant information such as date, time of observation and area. The period of observation was eighty-nine days, from 23 April to 20 July 1965. A block of ten days' worth of data, from 7 June to 17 June (days 46–56) was unusable because of camera malfunction.

Results were obtained in four runs, designated as Experiments 1, 2, 3 and 4. Experiment 1 was a preliminary observation period

in which it was assessed what the nature of the behaviour was; the other runs were tests for some notions arrived at on the basis of findings in the first period.

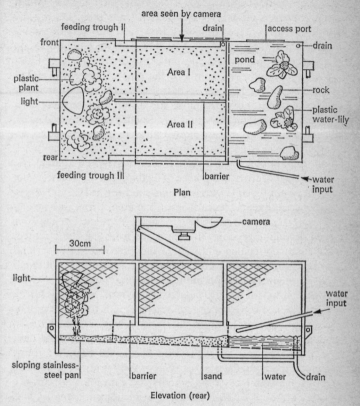

Figure 1 Plan and rear elevation of the frog terrarium. The light was an ordinary goose-neck desk fixture with a 100 W bulb

Results

The initial observation period was 23 April (day 1) to 10 May (day 18). During this period the existing feeding routine was maintained, i.e. feeding took place in Area I immediately following the frame taken at 16.00 h. Examination of the film footage disclosed

that the number of frogs present in Area I increased steadily during the day on *working days*, starting at about 09.00 h, reaching a maximum just after feeding, and declining rapidly again after working hours. On *holidays* (Saturday and Sunday) when the laboratory is closed, *no* such activity was seen. During the night few frogs were seen in Area I. In Area II only very occasionally an animal was present, night or day, working day or holiday.

In Figure 2 are shown the data gathered from the initial observation period. The points of the graph represent averages obtained by adding the numbers of frogs seen at any one particular time (such as 09.15) over the days, and dividing by the number of days. Eleven working days and five holidays are covered here. Only Area I observations are plotted; in Area II the average comes to less than 0·25 frogs at any time.

It is quite evident from Figure 2 that there was a gradual rise in the number of frogs in advance of feeding, but only on working days. At about 09.00 h the rise began, and toward 16.00 h there were between five and seven frogs in the area. From this we can extract a *response criterion*; it is reached if, in each of the four frames preceding feeding, five or more frogs can be counted. Immediately on feeding additional frogs emerged from the pond until there were ten to eleven frogs eating at the trough. (Although there were eighteen animals in the terrarium, I have never seen all of them eating at once. The maximum ever seen was fifteen animals; about twelve appears to be average for the entire experiment.) This number dropped quite rapidly after 18.00 h. On holidays on the other hand, there was no rise during the day; if anything, there was a slight drop during the morning hours, reaching a minimum of about 0·5 frogs near noon. There was a period of maximum activity ('activity' here meaning appearance on land) at about 02.00 h at night, a phenomenon that is also visible in the working-day data.

On the basis of these observations we can draw a tentative conclusion: the anticipatory feeding congregation of the frogs is a *conditioned* response. The unconditioned stimulus (US) is the appearance of food at a certain time and place. The conditioned stimulus (CS) is some function of the phenomenon 'working day' and the passage of time. The latter statement is supported both by the fact that start and termination of the response are strongly

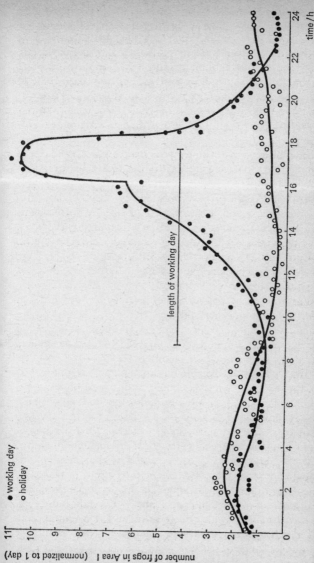

Figure 2 Frogs seen in Area I as a function of time of day. Dates from eleven working days and five holidays are represented, normalized to one day (see text). Feeding took place at 16.00 h on working days. Zero and 24 h correspond to 12 o'clock midnight, EDT

number of frogs in Area I   (normalized to 1 day)

time /h

length of working day

• working day
o holiday

correlated to the length of the working day (08.45–18.00 h), and by the fact that the response did not occur on holidays.

'Working day' is a complex phenomenon; it involves turning on of lights (the terrarium was in a rather dark corner of the lab, where only indirect daylight filtered through), and the usual noise and commotion of people at work. It would have been futile to attempt a complete analysis of relevant variables under the circumstances in which the observations were being made; an isolated space in which to continue the experiments was not available at the time.

There are, however, two simple tests that can be carried out to check if, indeed, the working day is the real CS (for it can reasonably be supposed that the frogs have sufficient memory to remember the alternation of five working days, two holidays, five working days, etc.), and to see if perhaps turning on the lights alone is a *sufficient* stimulus. For the first test an irregular holiday, falling in the middle of the week is required. If there were no response on this day, the 'periodicity' hypothesis can be rejected. Since there was no scheduled bank holiday in the offing at the time, we closed the laboratory on Wednesday, 5 May. The results were straightforward: there was no response and the frogs behaved exactly as they did on other holidays.

Having thus rejected the 'periodicity' hypothesis, the next test was carried out on the following Saturday (8 May). The lights in the laboratory were turned on quietly at 08.45 h, but no other activity was permitted. Again, no anticipatory response was evident. Light alone, therefore, is not a sufficient stimulus.

These initial observations on the anticipatory feeding response of the frogs allow us to formulate the hypothesis that the start of a working day is the CS, the response to which is the frogs' aggregation at the appropriate feeding trough, a certain amount of time after the CS. This hypothesis can be tested by changing the time of feeding, and/or the place of feeding; the pattern of responses should change to show *extinction* of the inappropriate response and *acquisition* of a new and proper one in a certain number of trials, *reacquisition* of a previously extinguished response, and, perhaps, *different rates* for learning a change of time, from those for learning a change of place.

Three experimental periods were planned for the test: Experi-

ment 2, from 11 May (day 18) to 24 May (day 32) during which feeding took place at the same spot as before (Area 1), but at 10.00 h; Experiment 3, in which feeding was again done at 16.00 h in Area I from 25 May (day 33) to 20 June (day 59); and Experiment 4, when feeding was done at 10.00 h in Area II for the remainder of the time, 21 June to 20 July (day 60 to 89).

Since the response is quite variable from day to day, and from sample to sample, averaging is necessary to reveal trends in behaviour. The data shown in Figure 2 represent a stationary situation, where averaging is done over time. The time-varying responses that will result from the changes in time and place of feeding, however, cannot be represented by this simple procedure. I therefore decided to use a running average, in which the observations in the appropriate time slots on five consecutive days were added together, and divided by five. The resulting values were then assigned to the third day. For obvious reasons, working days and holidays were strictly segregated although they were numbered consecutively. The arithmetic and book-keeping were done on a digital computer, which printed the averaged data not only in tabular form, but also in graphical form. The tabular form of the data is a 96×89 matrix, where the ninety-six rows represent the ninety-six film frames taken each day, and the eighty-nine columns the eighty-nine days of the experiment. At each row–column intersection is a number representing the number of frogs seen at that point in time. The running average is taken in the rows, and the result is that the first two and last two days of the experiment are lost.* This is due to the fact that the computer programme was designed to generate an average only if it had five consecutive inputs. One or two of these inputs could be 'no observation' (as opposed to the number zero); the computer would then divide the total of the five inputs by four or three, instead of five, for the average. In this manner, we obtain an automatic interpolation of missing data, provided not more than two datum points are missing in any sequence of five. The large

* Working days and holidays are separately processed; so we lose the first and last two days of each type. Since the experiment happened to start on a Friday and end on a Monday, *four* consecutive calendar days are actually lost at each end.

block of data missing from days 46 to 56 exceeds these limits, so a gap is left in the averaged data.

One important artefact, introduced by the averaging procedure is the rounding of sharp transients. Because the average for a particular day is obtained from the days that will *follow* it, as well as those that *precede* it, any sudden change is reflected backwards in time, and a response will appear to start two days before the stimulus is presented. Conversely, the size of a transient is decreased by the low-value data from the days preceding it.

If we imagine the averaged numbers to be stacks of unit cubes at the matrix points, we obtain a mountainous landscape whose peaks and valleys describe the behaviour of the frogs as a function of two units of time: hours of the day, and the succession of days. The best way to study such surfaces is by means of contour maps. Hence the computer was programmed to produce a set of maps. Each map is the hour–day matrix, but in the first the machine prints a mark at every matrix point where it encounters one or more frogs. The second map represents all points where two or more frogs are seen, the next map considers only three or more, and so forth, up to thirteen. Separate sets of maps were produced for Area I and Area II, for holidays and working days. These maps show patches of marks, which become smaller as the number of frogs represented by them become larger. By drawing lines around the perimeters of these patches and superimposing the outlines from each map in a set on one master chart, a proper contour map is obtained. Presumably one can programme the computer to do this task; however, it requires much less time and effort to do it by hand on a piece of tracing paper.

Figures 3 and 4 are the contour maps for Area I and II working days. Only the even-number contours and the five-frog response-criterion contour are drawn to keep the maps from being illegibly cluttered; for the same reason the lines have been drawn as if there were no holidays. In actuality, the landscape is slashed by trenches running from top to bottom on the plots every 6th and 7th day (and whenever an extra holiday occurred): these trenches go down to about the two-frog level.

It is evident in Figure 3 that there were three main areas of activity, corresponding to the feeding times of Experiments 1, 2 and 3. Also clearly visible is the fact that these areas of activity

were confined to the 09.00–18.00 h 'working-day' periods, with a fairly steep decline after about 18.00 h, and a gradual rise, except for Experiment 2, where the rise is steep. During the night-time hours rarely more than two frogs were out, except for the period around 02.00 h when, as we noted earlier, there was a slight

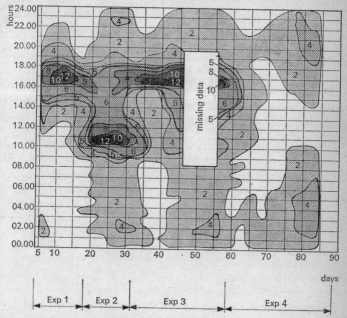

Figure 3 Numbers of frogs seen in the various stages of the experiment in Area I, plotted as a contour map. Increasing darkness of shading denotes increasing numbers, as indicated. Each number means that there were at least that many frogs present, but not as many as the next higher number. To avoid clutter, odd numbers have been omitted, except 5, which is response criterion. See text for details

maximum in the activity. No noticeable activity is visible during Experiment 4, which took place in Area II. In Figure 4, on the other hand, no activity is evident except during Experiment 4.

Detailed examination of Figure 3 shows the following. Immediately after the start of Experiment 2, the activity at 16.00–17.00 h began to decline. The 'filtering' effect of averaging, as we

noted before, makes it appear as if the decline started in advance of Experiment 2, just as the shift of activity to 10.00–11.00 h seemed to start in advance. This is, however, an artefact of the methods. Similarly, the transition from Experiment 2 to Experiment 3 shows this effect.

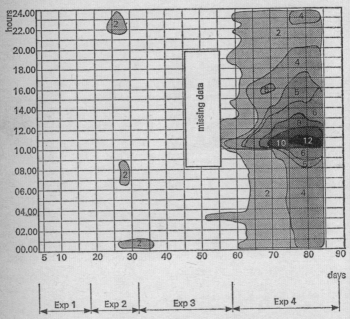

Figure 4 Numbers of frogs seen in the various stages of the experiment in Area II, plotted as a contour map. Increasing darkness of shading denotes increasing numbers, as indicated. Each number means that there were at least that many frogs present, but not as many as the next higher number. To avoid clutter, odd numbers have been omitted, except 5, which is response criterion. See text for details

During Experiment 1 the activity level rose gradually, as was clear from Figure 2 already (Figure 2 is tantamount to a profile through Figure 3 at about day 10), and exceeded criterion between 15.00 and 16.00 h. The extinction of this response is not clearly visible during Experiment 2, because after the feeding at 10.00 h the frogs remained in the area for most of the day. The establish-

ment of the new response in Experiment 2 is, however, quite clear. The criterion level of five frogs was reached about day 25, some six or seven trials after the start of the series. Again, the effect of averaging appears to push the response level back a few trials. By day 30 the response was fully established.

The transition from Experiment 2 to Experiment 3 (feeding at 16.00 h again) shows the extinction of the 10.00 h response quite clearly. With allowance, once more, for the averaging effects, the response extinguished in three or four trials. Re-establishing the 16.00 h response took a bit longer; some ten or twelve trials were needed (day 42 to 44) to reach the levels attained in Experiment 1. The extinction of the response after the start of Experiment 4 is also clearly visible and, again, was complete in three to four trials.

Figure 4 shows the establishment of the response to 10.00 h feeding in Area II. Although the acquisition pattern appears to be pretty much the same as in Experiments 2 and 3, the total time taken to reach criterion level was considerably longer: some twenty trials. It looks as if most of this was due to an initial latent period, running from day 58 to day 70, where no anticipatory activity was present at all. Even the number of frogs present *after* feeding did not reach the usual levels until about day 70. Then, from day 70 on, the response was acquired in some five to seven trials. A similar, though shorter, latent period was visible at the start of Experiment 3, in Figure 3.

Discussion

The hypothesis assumed after the preliminary observations of Experiment 1 stated that the start of the working day is the CS, to which the frogs respond by aggregating at the proper feeding trough, a certain amount of time after the CS. Testing this hypothesis by changing the time or the place of feeding should result in extinction of the inappropriate response, acquisition of a new and proper one in a limited number of trials, reacquisition of the previously learned response and, perhaps, different rates for learning time changes and place changes. The data of Experiments 2, 3 and 4 show that the hypothesis is essentially tenable. Extinction of one response and acquisition of another in relatively few trials is demonstrated, as is reacquisition of a previously

extinguished response. There is no distinguishable evidence that a change of place is learned at a different rate than a change of time, although the long latent period in Experiment 4 suggests that the frogs encounter more difficulty when confronted with a simultaneous change in place *and* time. The reason for failing to demonstrate the last point is quite obvious: too few experiments were made to see any trend emerge. One would have to run some half-dozen acquisition series to obtain enough data; but at the rate of one trial per day this would take several months.

The acquisitions and extinctions of the response shown in Figures 3 and 4 prove that frogs will learn and, in fact, do so surprisingly fast. One may wonder, then, why these experiments show such unambiguous results, while previous efforts to demonstrate conditioning in frogs have been largely unsuccessful. There appear to be at least two significant factors in which the present observations differ from previous ones. First, instead of forcing the frogs to perform some arbitrary task (such as jumping on a platform to avoid shock or collect a reward), I capitalized on a bit of natural behaviour, thus, in effect, leaving selection of the response to the frogs, but maintaining control of the stimulus. Second, instead of using a single individual, I used a group of frogs. Even with eighteen animals the response was variable enough to require averaging of the data; with a single animal the signal-to-noise ratio would likely have been so low that no reliable response could have been extracted, even over a very long time. On the other hand, working with a group of animals introduces a complicating factor: inter-individual stimulation. The experiments, unfortunately, give no indication of the strength of such an interaction, if it exists.

The results of this study contribute to our understanding of the frog's ability to survive, by demonstrating that frogs can learn to adjust their feeding behaviour to the time and place where food is available. Insects or other prey vary in abundance throughout the day and at different times of the year, and are found in widely varying locations. Unless the frogs have the ability to learn, and learn quickly, where food is most abundant at what time, and, moreover, have the ability to forget rapidly those places and times where the supply is exhausted, their chances of survival would be slim. The frog is a 'passive' hunter; it waits quietly until prey

moves within striking distance and then pounces. Unless the frog positions itself in a place where the probability of prey chancing within striking distance is maximal, it will likely starve to death. The results of this study demonstrate that frogs have indeed the capacity to adapt rapidly to their ever-changing environment.

W. M. Court Brown

The Study of Human Sex-Chromosome Abnormalities with Particular Reference to Intelligence and Behaviour

W. M. Court Brown, 'The study of human sex chromosome abnormalities with particular reference to intelligence and behaviour', *Advancement of Science*, June 1968, pp. 390–97.

During the ten years since the first description of an abnormal complement of sex chromosomes in a human male we have come to appreciate that some such complements can be associated with aberrant behaviour. In fact some of these abnormal males are repeatedly charged and convicted of crimes against the person or against property. We know also that many types of abnormal complement can be associated with diminished intelligence, an effect to be reckoned with in considering their effect on behaviour.

Comparative studies

The foundation of our knowledge of the influence of abnormal complements of sex chromosomes on intelligence and behaviour rests on comparative studies of the frequency of abnormal individuals in different groups of the population. In essence most studies to date have been on males and are based on a comparison of the frequency of abnormal males in the liveborn baby population with that of males among those in schools for the educationally subnormal or in hospitals for the mentally subnormal or in maximum-security hospitals or in prisons or in various training schools for young delinquents among which we may class the British Borstals, Young Offender's Institutions and Detention Centres. If, in statistical terms, there is a significantly greater frequency of individuals with an abnormal sex-chromosome complement in, for example, a group of mentally subnormal patients or among prisoners, then we conclude that there is a real and meaningful association present which reflects on why the relevant subnormal patients are in hospital or why the relevant prisoners are in prison. For example, the frequency of XXY males at birth is about 1·4 per 1000 liveborn males while among male

patients in hospitals for the mentally subnormal it is about 15 to 20 per 1000 for those who are classified as high-grade defectives with IQs of fifty or more. The difference is highly significant and we conclude that the presence of an extra X chromosome in these men has had very real influence on them so that they have come to be hospitalized in this way. Although our information on the frequency of XYY males at birth is much less than that on XXY males, we may for the present assume that their frequency is about 1 per 1000 males. The finding of a frequency of about 30 per 1000 males in some maximum-security hospitals suggests very strongly that the effect of the extra Y chromosome is in some way the reason for these men being in such hospitals.

This particular approach has imperfections and the most serious is that the frequency of abnormal males in a group under consideration, for instance those in hospitals for the mentally subnormal, is not compared with the frequency of such males among those not in these hospitals but in the ordinary male population and with a comparable distribution of age. Thus if the mortality risks for say XXY males were higher during childhood than for normal XY males, then the comparison must lead to an underestimate of the effect of the additional X chromosome which leads to increased numbers of XXY males in hospitals for the mentally subnormal. Another imperfection, particularly relevant to reports from hospitals for the mentally subnormal, is that for practically all of these we are not provided with the age structure of the patients. Most, if not all, such hospitals carry their quota of child patients, but there is evidence to suggest that many of the XXY males in these hospitals are not admitted until after puberty and often not until they are adults and often via the courts. It follows then that if we were able to compare the frequency among adults of XXY males in hospitals for the mentally subnormal with that among adults in the ordinary population, then the increase in frequency in the former could well be greater than shown by a comparison, as of now, between a mixed population of adults and children with liveborn babies. In spite of these drawbacks to our present methods of study we can at least say that our current estimates of the increases in frequency registered are more likely to be underestimates than overestimates.

The question of why the appropriate comparisons have not

been made taking account of age structure is an obvious one to ask. The answer is simple and it is that the laboratory techniques are time-consuming, and at present no laboratory in the world has the necessary capability to undertake studies of the ordinary population on a sufficiently extensive scale. To date most effort has gone into the examination of the liveborn baby population, but for the future it will be necessary to obtain information on age-specific frequencies. Thus, for example, one would like to know the frequency of XXY males per 1000 males not only at birth but by single years thereafter up to the age of five, and then by five- or ten-year age groups for the rest of the population.

Laboratory techniques

All these studies primarily depend on the exercise of two techniques, nuclear sexing and chromosome analysis. Nuclear sex is studied in cells which are not in division and the method is simple. It was discovered a good many years before the development of the modern techniques for preparing cells for chromosome studies that the cells of normal females show a well-defined intranuclear body in suitably stained preparations about 1 μm in diameter which is not present in normal males; this is the sex chromatin body. This discovery was made by Barr and Bertram in 1949 from the study of the neurones of cats. The sex-chromosome complement of normal females is XX and that of normal males is XY. However, the X chromosomes carry a number of essential genes which have nothing to do with sex determination, for example those concerned with colour vision and with some aspects of blood clotting. If both the X chromosomes of the female were functionally active then she would have double the quantity of functional X-borne genes than the male. The evidence is that this difference in gene dosage is compensated for by one of the female X chromosomes being inactivated, presumably during embryogenesis, and it is clear that this inactivated chromosome forms the sex chromatin body present in the nuclei of normal females but not in that of normal males. It also follows that males with an abnormal XXY complement demonstrate a sex chromatin body and are said to be chromatin-positive as in normal females. Furthermore, males with an XXXY complement or females with an XXX complement show two sex chromatin bodies, and, going

to more extreme abnormalities, males with an XXXXY complement and females with an XXXX complement show three sex chromatin bodies. In other words all X chromosomes in a cell in excess of one are inactivated and can be seen as sex chromatin bodies in non-dividing cells.

Nuclear sex is most conveniently examined in cells obtained by lightly scraping the inner aspect of the cheek, transferring the scrapings on to a glass slide, fixing the smear of cells and staining them. Such buccal smears are easily made in large numbers and this is a most convenient way of examining a group of males or females and identifying those individuals with an abnormal number of X chromosomes. It is to be noted, however, that nuclear sexing cannot provide information on the number of Y chromosomes, so that while a male with an XXYY sex chromosome complement will be identified as having two X chromosomes by nuclear sexing, this technique does not distinguish him from an XXY male. Similarly, nuclear sexing does not distinguish between the chromatin-negative female with only a single X chromosome and no other sex chromosome and the chromatin-negative female with an XY complement. The latter paradoxical situation is a feature of certain abnormal states of sexual differentiation. It follows, therefore, that once an individual has been identified with an abnormal nuclear sex, then chromosome studies are necessary to determine his or her sex-chromosome complement.

Nuclear sexing has another limitation. Quite a number of individuals are known who are chromosome mosaics, that is they are composed of two and sometimes more than two lines of cells which differ in their chromosome complements, and in sex-chromosome mosaics in their sex-chromosome complements. It is theoretically possible that some such persons may not be recognized in nuclear-sexing surveys if they are mosaics in which one line of cells has a normal male or female complement, for example XY/XXY or XO/XX mosaics. A study of mosaics has shown that there can be wide differences in the relative representations of the constituent cell lines in different tissues. For example, it is possible to identify a male who is chromatin-positive on examination of his buccal cells, but in whom a study of the chromosomes in blood cells or skin cells shows only a normal male complement. One would also expect individuals with things

ordered the other way round, and who would not, therefore, be detected by nuclear sexing. On the face of it the question might be asked: why, in such a person, are the chromosomes of the buccal cells not examined? The answer is that it is not possible to do this at present, and that for chromosome studies we are restricted to examining lymphocytes cultured from blood, fibroblasts cultured from any source, but most conveniently from a sample of skin, and the cells of the bone marrow which do not require to be cultured. From studies of nuclear sex in parallel with chromosome analysis on male babies it seems, however, that it will be only rarely that an XY/XXY mosaic will be overlooked by nuclear sexing alone, but there are not yet any comparable data for females. In spite of these limitations nuclear sexing has provided most valuable information on the frequency of individuals with an abnormal complement of X chromosomes and, combined with chromosome studies, on frequencies of males with XXY, XY/XXY, XXYY, XXXY and XXXXY complements.

When we come to consider XYY males then clearly nuclear sexing is uninformative, and the detection of these males primarily depends on chromosome studies. To determine their frequency at birth it is necessary to do chromosome studies on very large numbers of newborn babies, and the likely significance of the extra Y chromosome in terms of its effect on behaviour was only appreciated following the survey of the chromosomes of many men in a maximum-security hospital. Chromosome studies are time consuming and tedious by comparison with those by nuclear sexing. The cells of choice are lymphocytes cultured from blood samples in the presence of phytohaemagglutinin, which has the property of stimulating lymphocytes to increase in size and divide in the the conditions of culture at 37 °C. The cell shown in the top half of Figure 1 is one from a lymphocyte culture. It is feasible to have chromosome preparations ready for analysis within three to four days of obtaining a blood sample, which is not possible with the culture of fibroblasts from a sample of skin when it can take several weeks before preparations are available. The study of marrow cells is excluded for survey work as it is unethical to ask a healthy individual to submit to the discomfort of the procedure for obtaining bone marrow and one not entirely devoid of risk.

Micro-techniques have been developed for culturing lympho-

cytes and no more than 0·5 cm³ of blood is required. The development of these techniques has made possible the routine examination of the newborn, for the blood can be obtained from a skin puncture with no risk to the child and very little discomfort. The frequency of babies with an abnormal complement of chromosomes is about 1 per cent and the serious nature of some of these

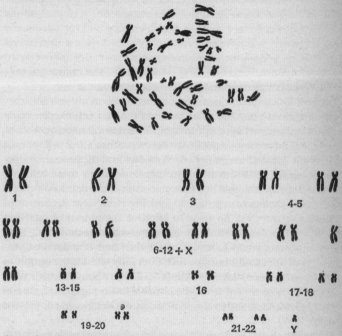

Figure 1 Chromosomes from a normal male

abnormalities, in terms of their effects on physical and mental development, provides adequate justification for the routine screening of newborn children. However, this can only be done on a limited scale at present for the bar to the development of such screening on a wide scale is the effort required for the counting and analysis of the chromosomes in a cell.

The top half of Figure 1 shows, considerably enlarged, the microscopist's view of a cell prepared for the examination of its

chromosomes. Enumeration of the chromosomes is simple but once counted these have then to be sorted out according to their size and morphology, as demonstrated in the bottom half of Figure 1. It goes without saying that if this is done, as many do it, by photographing the cell and cutting out the chromosomes from an enlargement to facilitate sorting them into groups, then the process is so time consuming that it it ill-adapted to mass surveys. The skilled cytogeneticist, however, can count and analyse the chromosomes by eye and on the microscope in an adequately prepared cell without loss of accuracy. But this is a demanding procedure, and on the whole one cannot expect one individual to deal with more than about fifty cells per day. The number of cells examined per person under study depends on the nature of the study. For example, mass surveys can be done on two cells per person, given adequate rules to cover the first cell chosen being abnormal and provided one recognizes that this limited number of cells will only provide adequate data on those all of whose cells carry the same abnormality. At the other end of the scale it may be necessary to deal with up to two hundred or more cells per individual, for example in the examination of the damaging effects on chromosomes of exposure to ionizing radiations. A number of centres are now taking steps to improve the output of analysed cells by the development of computer-aided techniques. Thus developments are in hand in the Clinical and Population Cytogenetics Unit at Edinburgh which hopefully will lift the output of analysed cells per year from about 75 000 to about 500 000. With such an improvement it will be possible to do population studies on a greater scale and to eliminate many of the flaws in our present data.

The cell shown in Figure 1 is one from a normal male and it can be used to illustrate two important points about the sex chromosomes. The first is that the X chromosome is one of a large group of medium-sized chromosomes, numbered in Figure 1 as $6-12+$ X, which vary little in size with all showing much the same morphology. In fact the X chromosome cannot be distinguished from the fourteen autosomes in this group by visual inspection, and it follows that this group contains fifteen chromosomes in normal males and sixteen in normal females. When an extra chromosome is present in this group then it is inferred to be an X chromosome

if the nuclear sex is correspondingly abnormal. So if a male is chromatin positive and has forty-seven chromosomes including sixteen medium-sized ones he is presumed to have an XXY complement, or if a female is doubly chromatin positive, that is her cells have two sex chromatin bodies, and she also has seventeen medium-sized chromosomes, she is presumed to have an XXX sex-chromosome complement. The second point is that the Y chromosome is quite dissimilar from the X. It is much smaller and has a different morphology. It is, in fact, rather similar to two pairs of small autosomes, nos. 21 and 22, but fortunately there are features about the Y chromosome which permit the cytogeneticist to distinguish from these two autosomes with confidence.

Effects on the phenotype

We now come to the problem of the effect of abnormal complements of sex chromosomes on the physical and mental features of an individual, that is on his or her phenotype. We may generalize here and consider three broad classes of persons. The first are those with a chromosome number of forty-five due to the lack of a sex chromosome. Such have a single X chromosome and can be referred to as having an XO sex-chromosome complement. No individuals are known with a YO complement and there are good reasons for believing this complement to be incompatible with life. All individuals with an XO complement in all their cells develop as females, and all these females as adults show infantile sexual development, are of short stature (always below 1·52 m) and often have other abnormalities, for example congenital malformations of the heart and the presence of very prominent skin folds on either side of the neck (webbing of the neck). In individuals who are mosaics with one line having an XO complement the effect on their phenotype depends on the nature of the other line or lines, and, it is also presumed, on the relative proportions of the different lines in the gonads. For example, females with an XO/XX sex-chromosome complement may range from those showing all the features of an XO female to those with the normal features of an XX female. Similarly those with an XO/XXX complement may range in their features from those of the XO female to those of the XXX female. Where one cell line has a Y chromosome, XO/XY or XO/XYY individuals, then the sexual

phenotype may be male or female and for the XO/XY subjects some are known to be true hermaphrodites. There is little evidence for XO females suffering any particularly deleterious effects in terms of intelligence and behaviour, so that such females or mosaics with a female phenotype and with an XO line will not be considered further.

The second general class consists of males with an additional X chromosome or chromosomes, these including XXY, XXYY, XXXY and XXXXY males together with a variety of mosaics, for example XY/XXY, XX/XXY, XXY/XXXY and XXXY/ XXXXY mosaics. The commonest of all are XXY males, and all these show some phenotypic changes. At birth they usually appear quite normal but from puberty onwards they show testicular degeneration and often there are other associated features such as deficient facial hair, breast development, a female configuration of pubic hair and eunuchoid proportions. The additional X chromosome seems to limit their IQ and, as already noted, their frequency is significantly increased among the mentally subnormal. All these men appear to be sterile, and some undoubtedly show behavioural disorders. XY/XXY males may range in phenotype from that of an XY normal male to that of the XXY male. All other forms of abnormal complement in which more than two X chromosomes are present tend to show the above effects more severely than in the XXY male while the XXXXY male often shows underdevelopment of the penis and scrotum as well as testicular atrophy. In quite a number the ulnar and radial bones of the forearm are fused together at the site of the upper radio-ulnar joint. The XXYY male shows the features of the XXY male, but in addition he may be very tall and his behaviour can be further affected by the presence of the extra Y chromosome. Altogether, therefore, males with an extra X chromosome show quite serious effects on their physical phenotype, in fact much more serious than seems to be the case for individuals in the next group – males with a single X but an extra Y chromosome and females with an extra X chromosome.

There may be some parallels to be drawn between XXX females and XYY males. The present evidence suggests that the majority of XXX females are normally developed females who menstruate regularly and who are fertile. However, in a minority

there is some underdevelopment of the secondary sexual characteristics and menstruation may be irregular and infrequent. In fact, menstruation in some ceases within a short time of its onset. The picture at present of the XYY male is that, when recognized in surveys of prisons, maximum-security hospitals and hospitals for the mentally subnormal, these males appear to be sexually normally developed, although information on their fertility is scanty. A number of the cases identified fortuitously before the link with aberrant behaviour was understood showed undescended testes. It is just possible, though at present there is no concrete evidence, that undescended testes are a feature of some of these males and here perhaps there is some parallel with the small proportion of XXX females with clearly inadequate ovarian function.

IQ and behaviour

We can now turn to the particular problem of abnormal sex-chromosome complements, IQ and behaviour, and this discussion will deal solely with males. This is not to say that XXX females may not have a low IQ and show behavioural disturbances. In fact there is clear evidence for a significantly increased frequency of these females in hospitals for the mentally subnormal, but nothing like the attention has been paid to them that has been given to abnormal males and our knowledge of the XXX female remains rather scanty.

Ferguson-Smith showed in 1958 that there appeared to be an unusual frequency of chromatin-positive males among the hospitalized mentally subnormal. This work has been extended by many others and by about 1961–2 it had been recognized that on average the frequency of chromatin-positive males in hospitals for the mentally subnormal was about 9 per 1000 considering all levels of IQ among the patients. Court Brown has reviewed the available data up to the latter part of 1967, and from those studies in which the distribution of IQ is known in the surveyed patients it is evident that the frequency of chromatin-positive males is significantly lower among those with an IQ of less than fifty (6·4 per 1000) than among those with an IQ of fifty or more (17·6 per 1000). We have also to remember that a proportion of these patients are children and it is likely that the frequency of

...ive males among these will be lower than among ...that the frequencies quoted above might well be ...adults were concerned, certainly for those with an ...r more.

...above included information from Sweden reported *in* ...Hambert. The Swedish practice in dealing with the mentally subnormal is to accommodate them as far as possible within the general community. However, where behavioural disturbances and other factors prevent this then the patients are committed to hospitals for hard-to-manage mentally subnormal males. Hambert reports on the nuclear sexing of five such hospitals accommodating those with an IQ of fifty or more and one hospital containing those with an IQ of less than fifty. These data were incorporated in the estimates given above of the frequency of chromatin-positive males with IQs of less than fifty and fifty or above. The frequency of chromatin-positive males among those in the higher grade of IQ was about 20 per 1000 and in the lower grade about 5 per 1000. Many of these patients had criminal records, but the important point is that they were not selected for hospitalization primarily on the basis of such records, and therefore, these hospitals should not be considered as maximum-security hospitals in the sense of the term as used in Britain.

The Swedish experience underlines an important aspect of the handling of the mentally subnormal, and it is that a low IQ *per se* may not necessarily lead to hospitalization. However, a telling factor in considering whether a defective can be accommodated in the ordinary population or requires to be segregated is the nature of his behaviour. There are sound reasons for arguing that a low IQ is not necessarily linked with disordered behaviour. For example most mongols are severely subnormal but a feature of mongols is their pleasant behaviour and attitude towards others. Conversely very severe behavioural disturbances are a feature of those developing Huntingdon's chorea, due to a dominant gene, but there is no reason for believing that the gene affects intelligence. In 1960 a nuclear-sexing survey was done on 600 unselected patients classified as sexual psychopaths at the Atascadero State Hospital in California with six being found to be chromatin-positive yet only one of these was assessed as a high-grade defective.

In short, behaviour and intelligence can be shown to be dissociated and it is not correct to ascribe the aberrant behaviour of someone who is mentally subnormal entirely to his or her lowered intelligence, although it cannot be denied that intelligence may have some influence in the matter.

The two English special hospitals of Rampton and Moss Side were examined by Casey *et al.* who surveyed the nuclear sex of the patients. These are maximum-security hospitals dealing primarily with dangerous and aggressive criminals the great majority of whom are mentally subnormal, and practically all the patients had IQs of fifty or more. Over 900 men were examined and the frequency of chromatin-positive males was about 22 per 1000 or very similar to the findings of Hambert. Casey *et al.*, however, examined the chromosomes of all their abnormal males, which Hambert had not done, and found seven out of twenty-one to have an XXYY complement, the remainder having an XXY complement. This survey was actually done some years before its publication and the findings were made available to Patricia Jacobs of Edinburgh in early 1965. She thought the proportion of the chromatin-positive males with an XXYY complement was unusually high, and she postulated that a chromosome study of men in a maximum-security hospital might well show unusual numbers of males with an XYY complement.

Jacobs and her colleagues proceeded to survey the chromosome complement of the males in the Scottish State Hospital, another maximum-security hospital. The preliminary findings were made known in 1965 and the complete findings in 1968. Males with an XYY sex-chromosome complement formed about 3 per cent of all the males in the hospital. Furthermore the XYY males were shown on average to be about 0·15 m taller than the XY males in the hospital, and taking men of 1·83 m or more then about one in four had an XYY complement, a quite remarkable finding. Casey *et al.* then examined fifty men of 1·83 m or more from Moss Side and Rampton and found twelve to have an XYY complement, confirming the work of Jacobs and her colleagues. In addition, Casey *et al.* examined fifty men of 1·83 m or more from Broadmoor, a maximum-security hospital dealing with those not mentally subnormal, and found four to have an XYY complement.

As was noted earlier in this paper the standard procedure for

assessing the significance of such data is to compare the frequencies found with those in the newborn population. While we have a lot of data on chromatin-positive males among the newborn, we had no information in 1965 and 1966 on the frequency of XYY males, simply because no one had done a chromosome study on babies. The only exception was a small group of 266 randomly selected male babies examined in Edinburgh, none of whom had shown an abnormal sex-chromosome complement. However, it was argued that XYY males at birth were not likely to be as common as XXY males which, as noted, have a frequency among liveborn males of about 1·4 per 1000. The basis of the argument was that the error which led to the production of a conceptus with an XXY constitution could occur either at the first or second meiotic division in the mother or at the first meiotic division in the father, while the error producing a YY sperm could only occur at the second meiotic division in the father. Therefore it was argued, and is still argued, that XYY males at birth are likely to be less common than XXY males. There are no published data but so far in Edinburgh we have only found a single XYY male in the examination of about 1100 liveborn male children. Whatever frequency is finally established we may conclude that to find 3 per cent of males in a maximum-security hospital to have an XYY complement cannot be explained by the play of chance. It indicates a significant association between the chromosome constitution of these men and their presence in the maximum-security hospital.

Given evidence for such an association it would be incorrect to conclude without further study that the association was a primary one, that is that the abnormal genotype was directly linked with aberrant behaviour. There are possibly secondary associations to be considered. For example, the additional Y chromosome may lead to mental subnormality and an individual may commit anti-social acts because of a defective intelligence causing a basic failure to be able to decide between one course of action or another. Another possible secondary association would be with unusual stature, as some psychiatrists argue. Undoubtedly a significant proportion of XYY males are unusually tall and limited data so far suggest that about half adult XYY males, without physical abnormality, may be six feet or more, whereas

one would not anticipate that more than 10 per cent of the ordinary adult male population were 1·83 m or over. We already know from more recent work that occasionally a young male adolescent in his early teens may be seen at a psychiatric clinic because of aggressive and delinquent behaviour, be quite exceptionally tall for his age and have an XYY complement. Here we are faced with the problem of whether his delinquent behaviour is a reaction to his unusual stature, setting him apart from his peers, or whether both features are a primary function of his abnormal genotype.

Price and Whatmore examined the XYY males identified by Jacobs and her colleagues at the Scottish State Hospital and they came to the conclusion that the evidence in this group of men, all classified as psychopaths, was that aberrant behaviour in most had been a feature from early childhood but that the family environment of these men had, in fact, not been particularly unfavourable. For example, there was virtually no crime recorded for the sibs of these men in contrast to a considerable amount recorded for the sibs of a control group of XY males, also classified as psychopaths and drawn from the same hospital. Nor was there any evidence of any correlation between childhood behaviour disorders and unusual physical size. In fact all the data were consistent with a primary association between the abnormal genotype and disturbed behaviour. This is not to say, of course, that the XYY male brought up in an adverse family environment may not also be affected by this environment. For all we know these males might be unusually sensitive to such adverse circumstances. Nor does the argument exclude the possibility that excess height may not also be a stimulus to antisocial conduct. However, the main point to make is that the limited evidence to date suggests that there is a direct relationship between the abnormal genotype and disordered behaviour in these men. For the future it will be of considerable interest to compare behaviour in XYY, XXY and XXYY males. In the latter two forms of abnormality we have a severe effect of gonadal maturation producing testicular atrophy, and, as already noted, often other effects such as breast development, absent facial hair and a female configuration of pubic hair. Here it may be very difficult to decide to what extent delinquent

behaviour is a reaction to these physical disadvantages coupled with a comparatively low IQ, and to what extent it might be primarily linked to an abnormal genotype.

At present there is a great deal of work going on in many countries to establish the frequencies of XXY, XXYY, and XYY males in a variety of different subgroups of the population. Already there is a certain amount of evidence pointing to increased frequencies in a number of prison populations, while the XYY male is being found in hospitals for the mentally subnormal, particularly those concerned with behavioural disorders. It is too early yet to forecast the upshot of all this work, but what is important is that the exploration of the influence of abnormal complements of sex chromosomes on behaviour, especially the study of the XYY male, reawakens interest in the general problem of the interaction of genotype and environment in the development of human behavioural characteristics. This is no bad thing, for to date the influence of many psychiatrists and criminologists has resulted in too narrow a conception of the nature of the wellsprings of behaviour with too predominant an accent on environmentalism.

17 E. H. Hazelhoff and H. H. Evenhuis

Importance of the 'Counter-Current Principle' for
Oxygen Uptake in Fishes

E. H. Hazelhoff and H. H. Evenhuis, 'Importance of the
"counter-current principle" for oxygen uptake in fishes', *Nature*,
12 January 1952, p. 77.

According to van Dam, the high utilization of oxygen in fishes is
due, among other things, to the circumstance that, in the gills,
water and blood flow in opposite directions. The question rises
whether this hypothesis can be verified experimentally. The
direction of the blood-stream in the gill lamellae is not liable to
changes; that of the water current, however, can be reversed.

We carried out some experiments with tench of about 16 g,
narcotized with 0·7 per cent ethyl urethane until the movements
of the opercula had just stopped. The mouth of the fish was tied
to a piece of rubber tubing. We verified that no water could pass
between lips and tubing by suspending some India ink in the
water passing through the latter.

Water was pumped through the tubing by means of a small
pump, driven by a synchronous motor with a variable retarding
attachment. The direction of the water stream could be reversed
at any moment by starting the motor in the opposite direction.

At one side of the fish the gills were put out of action by
pressing the operculum against the body wall. The volume of
water driven along the gills of the other half of the body was
exactly known. At the beginning of each experiment we worked
with opposite currents, and regulated the velocity of the water
stream in such a way that the utilization of oxygen amounted to
50–70 per cent. Afterwards the direction of the water current was
reversed, and the utilization determined again. Repeatedly we
verified at a magnification of about twenty times that the abnormal direction of the water stream caused no observable
change in the position of the gill lamellae. Therefore, the results
of our experiments cannot be ascribed to a modification of the
area of contact between water and gills. Finally, in some cases,

E. H. Hazelhoff and H. H. Evenhuis 217

Table 1

Experiment Number	Weight of fish /g	Velocity of water stream /(cm³ s⁻¹)	Temperature /°C	Utilization of oxygen (per cent)		
				opposite direction	same direction	opposite direction
1	13·2	0·16	21	28	3; 5	17
2	19·3	0·16	18	88	28; 17	87
3	19·3	0·16	16	57	6; 5	76
4	19·3	0·16	16	75	18; 13	50
5	19·3	0·16	16	69	4	
6a*	19·3		16	88; 89; 92	4; 6; 5; 3	
6b						26; 31; 31; 33
6c						53; 47; 44; 42
6d						
7	16·1	0·16	18	36; 43	1; 5	
8	19·3	0·16	18	38; 34	6; 3	
9	16·1	0·16	18	54; 46	21; 4	
10	16·1	0·15	18	39; 31	6; 8	
11	16·1	0·15	18	50; 61	13; 16	
12	16·1	0·15	17	44; 45	19; 15	

* Experiments 6a, b, c and d were carried out on different days

we returned to opposite currents at the end of the experiment, and once more determined the utilization. The results of the experiments are summarized in the accompanying table.

The variation among individual experiments is considerable. Nevertheless, the difference in utilization during opposite currents and during parallel currents is quite evident. Averaging over all the experiments, we find for thirty-two cases of opposite currents a mean utilization of 51 per cent, and on the other hand for twenty-five cases of parallel currents a mean utilization of 9 per cent. In other words, these experiments show that the 'counter-current principle' is of high importance for the efficiency of the fish gill, and no doubt for that of other gills as well.

G. M. Hughes

How a Fish Extracts Oxygen from Water

G. M. Hughes, 'How a fish extracts oxygen from water', *New Scientist*, 10 August 1961, pp. 346–8.

Fishes soon die for lack of oxygen when they are removed from water; this is so well known that we speak of 'a fish out of water' to describe an inability to cope with a new environment. Yet in water there is only one-thirtieth the volume of oxygen contained in the same volume of air. In this article I shall discuss the mechanisms which, although they cannot make use of the rich supply of oxygen in the air, enable the fish to solve the not inconsiderable problem of extracting oxygen from water.

A 100 g river fish needs about five cubic centimetres of oxygen per hour when at rest, and up to three or four times this amount when normally active. Even if it were 100 per cent efficient at removing this oxygen, the fish would have to pass at least fifteen to thirty cubic centimetres of water across its respiratory surfaces each minute. To handle that volume of air would not be difficult, but with water far more work is required because the density of the medium is nearly a thousand times greater and a hundred times more viscous. One or two per cent of the oxygen intake of a man at rest is spent in work done by the respiratory muscles. The figure for a fish is likely to be many times greater. In addition, the rate of diffusion of oxygen is 300 000 times slower in water than it is in air.

How, then, does a fish overcome problems which appear so much greater than those facing terrestrial vertebrates; and why should it die under conditions which appear more favourable?

Part of the answer to these questions lies in the structure of the respiratory surfaces of the fish and in the nature of the flow across them. The gills are a very finely divided series of plates, presenting an enormous surface of contact to the water, which passes across them in a single direction, unlike the tidal flow of

the mammalian lung. When the fish is removed into air, the loss of support from the water, together with surface tension, render the effective surface area of the gills extremely small: the result in most cases is oxygen deficiency and death.

The total surface exposed to the respiratory current of water varies between fishes according to their activity. In very active fish such as mackerel it may be over 1000 square millimetres per gramme of body weight, which is equivalent to more than ten times the external body surface. A measure of the efficiency of the extraction mechanism is its ability to utilize up to 80 per cent of the oxygen contained in the water passing over the gills. The highest figure determined in man is about 25 per cent. Such efficient extraction by the fish can be explained by the so-called 'counter-current' relationship between the flow of the blood and of the water, and by the remarkable pumping mechanism which affords a continuous flow of water across the gills throughout the whole respiratory cycle.

Counter current between the blood and water

The counter-current principle operates in many places in animals where efficient exchange of dissolved materials or of heat is required between two fluids. Such systems have long been used by engineers in heat-exchange mechanisms. One of the first to appreciate its importance in animal physiology was van Dam, who in 1938 described its operation in fish gills. Here it ensures that blood which is about to leave the gills almost fully saturated with oxygen meets water entering with its full oxygen content, while blood deficient in oxygen entering the gills meets water from which most of the oxygen has already been removed. In this way there is always a greater tension of oxygen in the water than in the adjacent blood, so that oxygen continues to pass into the blood throughout its passage across the gill.

The effectiveness of this arrangement is shown by the sharp decrease in uptake of oxygen (from 51 per cent to 9 per cent) when the direction of the flow of water through the gills is reversed experimentally.

For maximal effectiveness of exchange it is also important that the two fluids come into close contact and that their rates of flow are adjusted to one another. The distance over which the oxygen

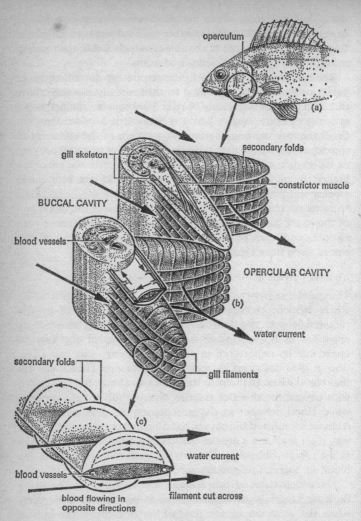

Figure 1 (a) Diagram to show the position of the four gill arches beneath the operculum on the left side of a fish. (b) Part of two of these gill arches are shown with the filaments of adjacent rows touching at their their tips. The blood vessels which carry the blood before and after its passage over the gills are shown. (c) Part of a single filament with three secondary folds on each side. The flow of blood is in the opposite direction to the water

has to diffuse from the water to the blood corpuscles is small, because the corpuscles are nearly as wide as the thin folds of the gill in which they circulate, and outside the folds the water passes on both sides (see Figure 1c). There is, as yet, little information concerning the volume of blood flowing through the gills in a given interval of time, but there are certainly reflex mechanisms which ensure that some relationship exists between the heartbeat and respiratory frequency. The heartbeat is usually slower than the respiratory frequency and in some instances it is synchronized with a particular phase of the respiratory cycle, but this is by no means always true. In a trout, for example, these frequencies are almost the same; they slip gradually out of step, although the heart tends to beat when the mouth is closing. In other cases the heartbeat is slower than that of the breathing in a simple ratio.

Such mechanisms ensure that there is always a good supply of water from which the blood can obtain its oxygen. This is important, because the blood can carry from ten to fifteen times as much oxygen as the same volume of water.

Continuous flow across the gills

When breathing, a fish opens its mouth and water is drawn into the buccal cavity; and after passing over the gills it leaves the opercular cavities through slits which appear when the gill covers (opercula) expand and come away from the side of the fish. This discontinuous flow into and out of the system gives a false impression of the flow across the gills themselves. Evidence for a truer description has recently been obtained by recording pressure changes on both sides of the gills by means of sensitive gauges – condenser manometers. These experiments on three freshwater species which I have done in collaboration with Dr G. Shelton, showed that with the exception of a brief period there is a pressure gradient (i.e. the pressure of water in the buccal cavity exceeds that in the opercular cavities) throughout the whole breathing cycle. It is almost certain therefore, that as a result water will pass continuously across the gills and so greatly increase the uptake of oxygen.

The mechanism which makes it possible involves the operation of two pumps slightly out of phase with one another, as illustrated diagrammatically in Figure 2. In the fish pumping actions are due

to changes in volume of the cavities, produced by muscular action. As we shall see later, the gill resistance is more complicated than the simple sieve depicted in the diagrams. During the inspiratory phase the buccal cavity expands and water enters

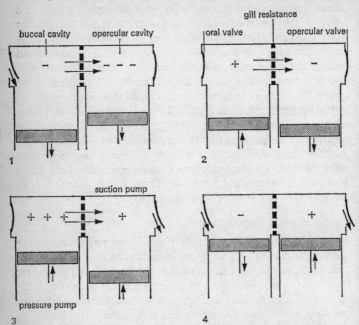

Figure 2 Diagrammatic representation of the double-pumping mechanism which maintains an almost continuous flow of water across the gills. Two major phases of the cycle are (1) in which the opercular suction pumps are active and (3) when the buccal pressure pump forces water across the gills. The two transition phases (2) and (4) each takes up only one-tenth of the whole cycle. The pressures in the cavities are given with respect to that of the water outside

through the mouth. At the same time the opercular cavities expand, but water cannot enter through their external openings because a thin membrane around the outer rim functions as a valve. During expansion of the opercular cavity the hydrostatic pressure becomes less than that in the buccal cavity and water is drawn across the gills: the opercular cavity acts as a *suction pump*.

During the phase of decrease in volume, which also starts in the buccal cavity, the pressure becomes greater than the external water as the mouth begins to close. It is functionally closed even in those fishes which do not close their mouths completely because of the presence of two thin membranous valves which project from just behind the upper and lower lips. During this phase, the increase in pressure within the buccal cavity is greater than that in the opercular cavities and water continues to pass across the gills: the buccal cavity acts as a *pressure pump*.

The pressure-difference curves (Figure 3) show that during nearly the whole of the cycle there is an excess of pressure tending to force water across the gills from buccal to opercular cavity. There is a brief period, however, during which this pressure difference is reversed and there will be a tendency for the flow to go in the opposite direction, but, as it is so brief and the pressure difference so low, the inertia of the water makes any actual reversal unlikely. In this way, then, a continuous flow of water is maintained over the gills in a direction opposite to that of the blood and a very high percentage of the oxygen in the inspired water is removed from it.

There are many interesting variations on this basic plan which have been revealed using similar techniques on marine species; for example, the relative contribution of the two pumps, buccal and opercular, in fishes occupying different habitats. Fish which are predominantly swimmers have the buccal pump better developed, although in some cases neither pump operates during the swimming movements of the fish. A notable example of this is the mackerel, which is obliged to swim continuously in order to maintain the flow of water through its gills. In other cases, such as leopard sharks which I observed in the Marineland Aquarium, California, pumping movements are not present during swimming, but appear as soon as the fish comes to rest.

Fish which spend most or all of their time on the sea bottom have enlarged opercular cavities supported by additional skeletal rays and the suction pump is better developed. Such fish as the bullhead, gurnard, dragonet, plaice and other flat-fishes are of this type. In the dragonet, for instance, gradual opercular expansion maintains a fairly constant differential pressure across the gills; then, during a relatively brief contraction phase, water

is expelled from both cavities out of the narrow opercular openings.

In flat-fishes which lie permanently on one side and are almost buried when resting on the sea bottom, other problems arise. The gills are equally developed on both sides of plaice and sole, and it seems certain that the water is pumped through both

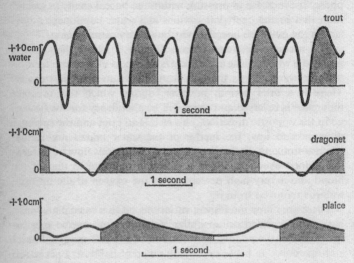

Figure 3 Curves to show the pressure difference between the buccal and opercular cavity in three fishes. Positive values indicate a greater pressure in the buccal cavity. The opercular suction phase (4) is shaded and forms a greater part of the cycle in the dragonet and plaice. The absence of any negative phase in the latter is probably because it can actively close the opercular opening

opercular cavities. The danger of sand entering the cavities and damaging the gills is a very real one. No reversal occurs in the differential pressure of these fish, and this is probably due to the active control of the opercular valves which will, therefore, prevent the entry of even the slightest current.

Just as the working of the two pumps is related to the habitat of the fish, so there are corresponding variations in the structure of their gills. Bottom-living fish generally have smaller gill areas and coarser sieves. The sieves are made up of many thousands of

pores which are found between the gill filaments – the double rows of thin plates which are stacked all the way round each of the four bony arches on both sides of the fish (Figure 1(a)). They form a continuous meshwork extending across the whole of the side walls of the pharynx. As the tips of the filaments are splayed out by the elastic properties of the supporting skeleton they remain in contact with one another at their tips, and water passes between the slits provided by the plates of adjacent filaments. These are not simple slits, however; the upper and lower surfaces of each filament have thin projections which form the actual respiratory surfaces. It is the collapse of these secondary folds and the consequent reduction in effective area for gaseous exchange that leads to asphyxiation when most fish are brought on to the land. The closer these folds (horse mackerel thirty-nine per millimetre, herring thirty-three), the more readily they occlude one another. In those fishes of the sea-shore that are exposed by the tide, such as some gobies, the secondary folds are widely spaced (fifteen per millimetre), and different species have supporting structures which are greater the longer their exposure on the sea-shore.

The sieve provided by the gills is a very fine one and is shown in Figure 4. At first sight the dimensions appear too small to allow sufficient flow of water with pressure differences of only one three-thousandth of an atmosphere which have been found in many species. The number of pores is so large, however (for example, a quarter of a million in a 130 g tench), that calculations based on an equation similar to Poiseuille's for the flow through each of them give figures for the total flow which are at least as great as the rates of flow which have been measured in fish at rest. But at higher rates it is likely that an increasing fraction of the water escapes between the tips of the filaments. Such 'short-circuiting' may account for the lower utilization of oxygen and the apparent fall in gill resistance which we have observed at these rates of flow. The gill resistance is certainly not constant during the respiratory cycle. Cine-films of small transparent gobies and young conger eels have shown that there is a definite phase in the respiratory cycle when the tips of the filaments separate and allow an increase in the short-circuited flow. During very active pumping, the process will be accentuated by

the yielding of the gills to the increased pressure difference. Contact between the filament tips is maintained entirely by the elasticity of the gill rays, there being no active muscles which can keep them spread out. The constrictor muscles shown in Figure 1(b) function when the fish makes 'coughing' movements,

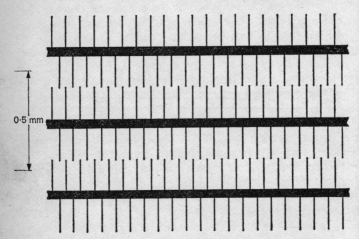

Figure 4 Diagram of a part of the sieve provided by the filaments and their secondary folds in a tench. The section passes through three filaments and shows secondary folds projecting alternately above and below the surface of each filament. The water flows at right angles to the paper

whereby the pressure gradient is reversed and a rapid reversal of flow cleans the gills.

These are but a few of the features found in the respiratory mechanisms of fishes during investigations of the past few years. A great deal remains to be learned about the detailed relationships between the flow of water and blood across the gills and the uptake of oxygen, to say nothing of the wide variety of adaptations which await investigation in the fish living under different conditions. Like so many problems investigated by present-day zoologists, the physical and physiological aspects have here illuminated our extensive knowledge of the animal's natural history and structure.

Part Three
Structures, Processes and Control in Populations

Populations behave very much like organisms; they grow, they die, they have internal systems of communication, and they appear to have homeostatic mechanisms which regulate their activities. The papers in this part are concerned with these properties of populations.

The means by which the size of a population is regulated are the subject of a controversy in modern biology. Many factors interact to determine this size, but probably the most important is the availability of food. If this is in short supply, the population must be reduced, by some sort of homeostatic mechanism. The debate is about the form which this takes. Reading 19 by Perrins presents some of the evidence for one of the viewpoints, and a critical discussion of their significance. He starts off with a knowledge of both of the major theories and gathers data purposefully. When he cannot gather the sort of data which would allow him to decide between the theories, he sets up an experiment which will provide the data required. In this case the experiment was to increase the brood-size for some birds by adding young of the appropriate age to a nest. The birds in each nest are marked with a small numbered ring attached to a leg which, when individuals are trapped, enables the brood in which they were brought up to be identified. The information gathered during the course of Perrins's study is that (a) birds which are heavier when fledged have a greater chance of survival, (b) birds from smaller clutches tend to be heavier, and (c) so do birds hatched earlier in the season. The contribution of each of these to the eventual survival of young is discussed, and then the two theories (that of Lack, and that of Wynne-Edwards) are considered in the light of Perrins's evidence.

Perrins comes to the conclusion that the evidence he has gathered tends to support one of these theories rather than the other. In the section headed 'Discussion', he assembles each individual piece of evidence into a logical argument. But the results of this investigation cannot be extrapolated to cover all species at all times. There is very good evidence that the other theory considered by Perrins holds for other species in different situations.

Human populations are subject to fluctuations, and Revelle discusses these variations in exactly the same terms as Perrins does with his great-tit populations: food supply, 'brood-size' and mortality rates. Does one of the two theories of population-size control mentioned by Perrins affect human populations, or is there something else involved in human-population control which is not found in the regulation of other animals' numbers? However, man is able to control some of the parameters which affect the size of his population, as will be discussed later.

In order to be able to control themselves, populations have systems of communication, as do individual organisms, and there are interesting parallels with the systems in individuals. Pheromones are the approximate equivalent, for populations, of hormones, and serve to integrate the activities of the members of the community. We are only just beginning to realize the important part they play in the regulation of the many communities of animals, vertebrate and invertebrate. Moore's article (Reading 21) shows how important they are in the life and evolution of social insects.

Signals emitted by one organism need not be chemical in origin, however. They may be vocal, or produced by special organs, or by the use of organs primarily developed for another purpose, as for example in the courtship behaviour of fruit flies, where the male signals to the female with his wings. Bennet-Clark and Ewing show how this behaviour differs between closely related species, and how it helps to prevent interbreeding between different species, thus preserving as separate entities the pools of genetic variability known as 'species'. In Reading 23 Harris provides an interesting

comparison of two closely related species of gull found around our coasts. Behavioural mechanisms also keep species separate in this case, but here there is evidence that the behavioural elements involved can be learned – the birds choose mates similar in appearance to foster-parents on which they have been fostered as eggs. Whether learned or innate, it is shown in both these papers that behaviour is very important in defining a species.

Homeostatic mechanisms exist which regulate genetic variability within a population. Sometimes, changes occur in the environment which prevent this equilibrium being maintained, and then changes in the frequency of certain genes takes place. (These changes are otherwise known as 'evolution'.) The study of gene frequencies in a population is called population genetics, and some fascinating work can be done with the study in schools. One such study is that by Jukes, when he was a sixth-former. He studied the frequency of 'tongue-rolling' in some populations in the Midlands and came up with some interesting ideas for further investigations. This work prompted further thoughts on the matter from Thompson in a later issue of the same journal.

When man was a hunter of animals and gatherer of plants he had to content himself with the animals and plants provided by the processes of natural selection. About 10 000 years ago he began to cultivate the land, and then a process of selection began which produced organisms more suited to his needs. He was unwittingly applying techniques which have now been analysed and improved by the science of genetics. Riley describes some of the results of these studies, which have produced organisms suited to man, rather than to the environment in which he found them.

Human populations are subject to natural selection in the same way as any other population, but in addition they can also adjust the environment to their requirements. This is of increasing importance. The remaining three papers are all concerned with the way man manipulates his natural environment in order to increase his food supply. Tannenbaum and Mateles write about the attempts being made to utilize

micro-organisms for food, and Bowers shows how biologists are improving protein supplies from fisheries. Finally Knipling speculates about the possibilities and advantages of using sterile males to reduce the fecundity of insect pests. Since 1955 when his paper was written, the technique has been used with considerable success in several parts of the world.

19 C. M. Perrins

Population Fluctuations and Clutch-Size in the Great Tit

Excerpts from C. M. Perrins, 'Population fluctuations and clutch-size in the great tit', *Journal of Animal Ecology*, vol. 34, 1965, pp. 601–47.

[This paper describes experiments carried out in Marley Wood, at Wytham, near Oxford. A number of nest-boxes were put up at a density of about three per acre, to provide artificial nest sites for great and blue tits.

The first part of the paper has been largely omitted from these extracts, and deals with the population of great tits at Wytham, and their movements and survival. A large section is devoted to the way in which the timing of the breeding season is related to availability of food, the spring temperature, the age of the birds (older ones lay earlier than younger ones) and the type of habitat.]

Methods

Each nest-box was visited at least once a week during the nesting season; since the great tit lays one egg per day, it was possible from such visits to get the date when each clutch was started and completed. Every great-tit nest was visited each day from the eleventh day after completion of the clutch until the first young hatched. Thus the date when each brood hatched was known.

On the day of hatching some of the broods were artificially altered in size. For example, if two broods of nine young hatched, three young might be transferred from one brood to the other, making one brood of six and another of twelve. The down of transferred young was dyed with gentian violet, which was normally visible for at least ten days. This was done mainly to ascertain whether the foster parents discriminated in any way against the strange chicks; they did not. The reason for manipulating the size of broods was that when studying survival of the young in broods of different sizes, Lack, Gibb and Owen had

very few data on the broods of above the normal size. To give some idea of the extent of this operation, some half to one-third of the broods were used for manipulation (excluding late broods); virtually all broods of eleven young or more were artificial.

The young birds were weighed on the fifteenth day after hatching and their weights were recorded against their ring numbers. Thereafter the nests were not visited until well after the time at which the young should have flown (about the nineteenth to twentieth day after hatching) to avoid any possibility of causing the young to leave the nest prematurely. It was nearly always possible at the final visit to ascertain whether the young had fledged successfully or whether they had been attacked by a predator.

[The next part, extracts from which are included here, deals with the size of a brood, and the way this is related to the survival of the young. The term 'recovery' relates to a bird which is re-trapped at a date after it had first been marked with a ring.]

Clutch-size

While several factors will be shown [not included] to affect the size of the great tit's clutch to varying extents, the average clutch in Wytham is usually about nine or ten. Lack developed the theory that the clutch-size of birds was adapted to the largest number of young which could be raised successfully; Lack gave some evidence that this was true for the great tit. It is now possible to examine the situation further.

Gibb showed that the larger the brood the more often the parents feed the young, but they are not able to increase the rate of feeding in proportion to the increase in the number of young. 'A member of a small family therefore received more food per day than did one of a larger family.' This situation has been recorded for many species of bird. As might be expected, the result is that the young of the large broods are lighter in weight; Figure 1 shows the range. (It will be shown below [not included] that young in later broods are lighter than those in early ones and that brood-size decreases with season. Hence large broods occur most commonly when young tend to be heaviest and so the weights in this table are biased in favour of the large broods and

the difference, at any given time of the season, is actually greater than that shown.)

In Wytham, omitting nests destroyed by predators, 90–95 per cent of the young leave the nest successfully, regardless of their weight. Many of these young are caught [in traps or in special nets] during the following winter and analysis of these in relation

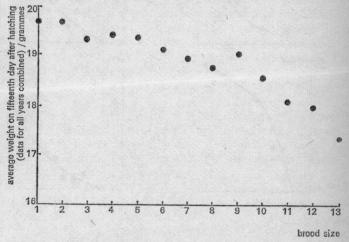

Figure 1 Great tit: weight in relation to brood size

to their weight (Figure 2) shows that more of the heavy than of the light young survive. Hence proportionately more of the young from the smaller broods survive than those from the large broods.

Figure 2 shows the recoveries in terms of the number of surviving young per brood (the important measure from the evolutionary point of view). The large broods do not always produce proportionately more surviving young than those of normal size. Figure 2 shows the survival rate in relation to brood-size for the blue tit, where the same holds true. The figures given here include broods that were taken in the nest by predators since predators tend to take larger broods (Figure 5). [. . .]

There is little doubt that the large broods are taken more

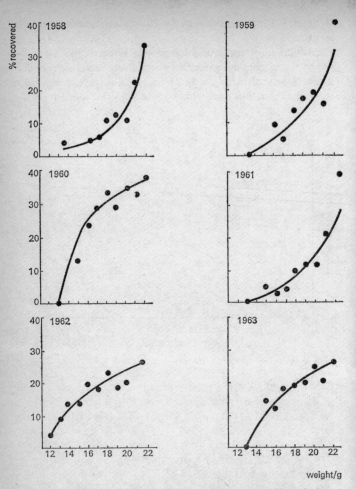

weight/g

Figure 2 Survival of great tits in relation to weight on the fifteenth day. Any bird which was known to be alive three months after fledging was counted as having survived. Owing to the very small number of recoverings of birds of 15 g or less, in most years information for the birds of 12–15 g has been lumped and presented as one point. Since the weights of the young form a normal distribution, data on the very light and the very heavy are fewer than those of the more average weights. The lines are drawn through the points by eye

frequently by predators because they are hungrier and calling more.

In addition to the loss of the young, about 20 per cent of the females seem to be taken with their broods, thus increasing the disadvantage of raising a large brood, since these females will not get a further opportunity to breed.

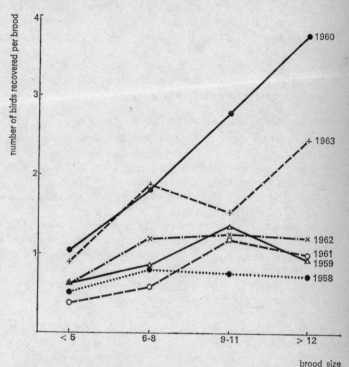

Figure 3 Recoveries of great and blue tits in relation to brood size

Figure 3 is biased in favour of the large broods. The great majority of the broods of eleven or more in the great tit are artificial in that, as mentioned earlier, the broods were increased by three or more young at hatching. These broods appear to be reared in a perfectly normal way by the parents. However, they differ from natural broods of the same size in that, had the

parents laid a clutch of that size, they would either have had to start laying earlier (for which they might not have had enough food) or the clutch would have been completed later, and consequently hatched later, by three or four days. Such delay would have been accompanied by lower survival [evidence not included in this Reading] since the date of hatching markedly affects the survival of the young. Hence the young in these artificially increased broods, had they been natural, would have hatched later and survived less well than the figures show by perhaps 15 per cent.

Hence the commonest brood-size is that which normally produces the most surviving young. A larger brood results, on average, in fewer rather than more surviving young.

Factors regulating total population change

It is clear that the major factor affecting the numbers of breeding tits is the survival of the young in the previous year; survival of adults and movements of birds will modify the changes. Movement does not seem to be important in Wytham.

Kluijver has shown that adult mortality is much less varied than that of the juveniles in this study.

It is the survival of the young birds after fledging that varies so markedly and has considerable effect on the population [as shown in Figure 4]. Kluijver has shown that the production of young is partly dependent on the density of the parents; the lower the number of breeding pairs the higher is the number of young raised per pair. The same is also true in Wytham. Some people have suggested that such behaviour is designed to reduce overpopulation. Two points suggest that this is not true for the great tit. Firstly, the reduction in brood-size is not nearly proportional to the increase in density. Secondly, the actual production of young to fledging is not closely correlated with the number surviving until the autumn. Indeed there is no significant correlation between the number of young fledged and the change in the following year's breeding population.

The main point about the birds that disappear in this study is that they are known to have been physically inferior to the survivors in that they were lighter, either by virtue of their brood-size or because they hatched at a later date. They are therefore

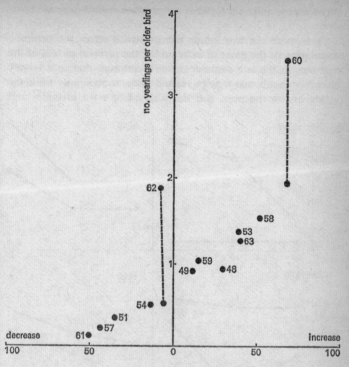

Figure 4 Great tit: ratio of young birds to older ones in relation to percentage change in the breeding population the following year. The percentage change was calculated as follows: the difference between the numbers breeding in year A and year B was calculated as a percentage of the number in year A. If the population was larger in year B than year A then the percentage was plotted as an increase: if it was smaller as a decrease. In most years there did not seem to be much change between the ratio at the beginning of the autumn and at the end of the winter. This was not so in 1962 when there was an apparently very high mortality of juveniles during the very hard winter. In 1960 virtually all the birds were caught under beech trees, feeding on the seed. More juveniles seem to wander in the winter and come to the beech areas. Thus the number of juveniles is disproportionately high (in fact impossibly so since the birds did not raise seven young per pair in the previous summer). The ratio on a small sample trapped in Marley (i.e. not under the beeches) was only 2·0 young per older bird and there was 1·8 young per older bird breeding in 1961. Hence for 1960 and 1962 two points are shown, that for the high winter ratio and that for the spring

those that would be less likely to survive a period of food shortage.

Thus there are two major factors which affect the breeding population of the great tit in Wytham, the survival or loss of the juveniles in the late summer or early autumn, due to a largely unknown factor that may be related to the food supply while the young are in the nest, and the survival of both juveniles and

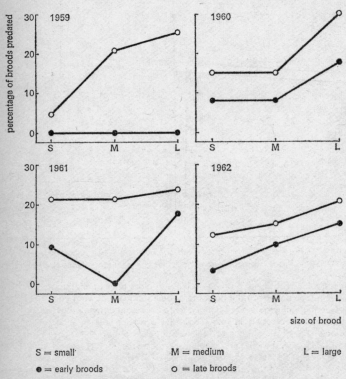

Figure 5 Predation on great-tit nests in Wytham in relation to brood-size and time of season, 1959–62. Each year is subdivided so that approximately half the data will fall into each half of the season. Similarly brood-sizes are chosen so that similar numbers of nests will fall into small, medium and large broods. The data are biased in favour of large broods, as the very late broods suffer the heaviest predation and most of these broods are small

adults through the winter which depends to a large extent on whether there is a beech crop or not, the mortality being higher in years without a crop. [As shown in Figure 6.]

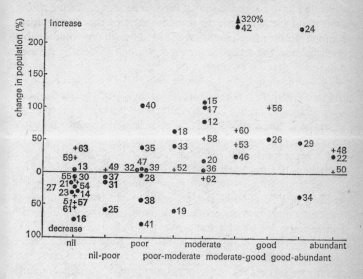

Figure 6 Great tit: changes in breeding population in relation to the crop of beech-mast. The data for the Dutch populations are from Wolda and Kluijver. The beech crops are given in the categories listed by the Forestry Commission. They are subjective and probably the increase between one category and the next does not represent a constant increase in the quantity of seed. The percentage change in the tit population was calculated in the same way as in Figure 4. +, Wytham 1948–63; •, Holland 1912–42, 1946–7

Discussion

Selective factors affecting breeding

[. . .] There is great advantage in being an early breeder, but [. . .] many of the great tits do not breed early, and this can only be because something prevents them. The most likely factor seems to be that the birds are unable to get sufficient food to manufacture eggs earlier. It will be remembered the older birds breed earlier than yearlings, perhaps because, being more experienced,

they are able to get enough food earlier in the season. The food taken is very varied at this time, but composed mainly of small invertebrates.

While I have suggested that at least some element of the spring food is sufficiently scarce to prevent many of the females from getting into breeding condition at the time which would enable them to raise the largest number of young, it must be stressed that the evidence is circumstantial.

The ultimate factor to which the tits' breeding season is adapted is the food supply for the young, which leave the nest, on average, some 10–12 days after the peak of winter-moth abundance. Since those that leave earlier survive much better than those that leave later it is evidently important to leave the nest when food is still plentiful. Yet many of the young leave later than the time when food is most abundant.

Interesting support for this suggestion that the female great tit finds it difficult to collect enough food to manufacture eggs comes from Royama's work in Japan; he observed that the female starts begging for food from the male as soon as she has started laying. From this time onwards the male feeds the female about five times an hour throughout the laying period (i.e. some seventy meals per day and about 30 per cent of the female's daily food). Such feeding has been called courtship feeding, but one would expect selection to have established the habit of the male trying to help the female obtain sufficient food if it is in short supply.

Hence, although there is presumably selective pressure towards breeding at the best time in relation to the caterpillars, the time at which they actually get into breeding condition is not related to the caterpillars' season but to the supply of spring food. (The supply of spring food, like the timing of the caterpillar season, is probably largely affected by the weather. This would explain why the tits' breeding season is approximately correlated to the time of caterpillar abundance since both the spring food supply and the caterpillar season are likely to be similarly affected by the spring weather.)

Many points remain to be elucidated before this suggestion can be considered proved but there is, perhaps, one main reason for treating the idea with caution. This is that many other species

vary their clutch in the same way as the great tit, the first clutches being largest and a steady decrease in clutch-size occurring throughout the season. It seems likely that, as in the great tit, the largest clutches are laid at the time at which they can be most easily reared, and that selection has favoured the laying of smaller clutches later in the season because it is not possible to rear quite such large broods at this time. Thus the first birds to lay rear the largest number of young and the others would be more successful if they laid earlier.

In conclusion it is clear that the great tit shows a remarkable series of responses to the external conditions when laying its clutch, although the mechanism by which it does so is not understood. The birds are not able to forecast the food supply directly and their responses are based only on the conditions normal to a given set of circumstances and, while they lay a size of clutch that is reasonably well adapted to local conditions, they sometimes respond inappropriately when these are exceptional. In addition, since conditions for rearing young vary so much from year to year, the most productive clutch-size varies also. Because of this a fairly wide range of clutch-sizes is still found in any population, presumably because each has been the most productive too frequently for it to have been eliminated by natural selection.

Survival of birds after leaving the nest

Some young great tits are more likely than others to survive after leaving the nest, but it has not been possible to show exactly how the mortality occurs. One of the most striking features that has emerged is the great variation in post-fledging survival in different years, the ratio of young to older birds varying in different winters between 0·16 and 2·0, and the mortality varying between about 90 per cent and 20 per cent or less.

It has been suggested that the heavier young may be carrying more fat than the lighter ones and that it is largely this food store which enables these young to survive better than the lighter ones. In addition, among birds of equal weight, survival was better from the early part of the season than from the later part.

It seems likely that a young bird will stand a much better chance of surviving this critical period in its life if it has an internal food reserve, on which it can survive for a day or two in

an emergency, than if it has no food reserve, particularly later in the season.

It seems extremely unlikely that many more of the light, rather than the heavy, young would die from predators or disease until they had become sufficiently weakened by starvation to be particularly susceptible. In other words, much of the predation which occurs after the young have left the nest can probably be looked upon as a side effect of food shortage in the same way as can much of the predation in the nest.

Extra fat on heavy chicks also conveys extra advantage in the form of insulation on cold nights. This seems very unlikely, however, to be of adaptive significance since the period when the young are first out of the nest is in late May or June when the nights are not very cold. Also, the extra weight of the heavy young seems to be of especial advantage to the late young and not the early ones, when it is even warmer.

Another aspect of the survival of the young birds which needs investigation is how well the young in the poorer habitats – gardens and pinewoods – survive after leaving the nest. Their low weights make one doubt whether many of them live long after leaving the nest – in most years at any rate. In Wytham, except for 1960 and 1962, most of the young of these weights did not survive. However, it is possible that the conditions after the young have left the nest are not so bad in gardens as in oak woodland; and the same might also apply in pinewoods where a reasonable food supply is present for much longer than in oak woodland.

Snow found that blackbirds, *Turdus merula*, unlike tits, were both denser and had better breeding success in gardens than in woodland, but here the main difference in survival between the two habitats was the heavier predation in woodland.

Conclusion

Currently, there are two major theories on the significance of reproductive rates and their relationship to adult mortality. The distinction has been clearly made by von Haartman (1954). One theory is that animals are reproducing as rapidly as they can and, in the case of birds, that birds lay a clutch of such a size that the parents raise the largest possible number of young. By doing so

these birds leave the most progeny and their genotype becomes the most common in the population. This von Haartman calls the theory of inter-individual selection. The other theory is that a certain number of offspring is needed to compensate for the average mortality of the species. If a group of birds produce more young than is necessary then there will be too many birds, food will become scarce and that population will become extinct. It will be replaced by another population whose individuals, by laying smaller clutches, can maintain an optimum density. This, von Haartman, like others, calls the theory of inter-group selection.

One of the critical differences between these two theories is that in the first the birds' numbers are assumed to be limited by food outside the breeding season while, in the second, the birds would starve if there were more of them, but the birds lay fewer eggs to prevent this happening. The proponents of both theories agree that birds are, ultimately, limited by their food supply, but those holding the first consider that the individual animals are in direct competition for food, while those holding the second believe that the animals have evolved mechanisms which prevent their numbers reaching a level at which any direct competition for food is necessary.

I am of the opinion that the Wytham great-tit population shows some tendencies which support the first theory and very few which support the second. Firstly, it is difficult to believe that the tits are not breeding as fast as they can. This does not mean that, in every year, they start with the largest number of young that they could possibly raise, since in some years it seems likely, in retrospect, that they could have raised more than they did. However, it seems probable that the birds are laying a clutch from which they will rear the largest number of healthy young under the average conditions that are found when the young require feeding. There is also strong evidence that the clutch-sizes are adaptively adjusted to various conditions such as habitat, date of laying, density of breeding pairs and age of the individual birds. But some variations do not appear to be adaptive, such as the laying of smaller clutches in a late season, although the tits may not be able to avoid this. Similarly in Britain the great tit in pinewoods responds inappropriately because it is adapted to broad-leaved, deciduous woodland.

In broad-leaved woodland there is clear evidence that in some years the parents of the largest broods do not produce as many surviving young as those birds whose broods are slightly smaller. Thus if the tits had larger broods they would raise fewer, rather than more, young.

Another way in which it would be possible for the tits to raise more young would be to produce more than one brood. This, in oak woodland, they do not normally do. In this paper attention has been drawn to the fact that the birds do not breed as early as perhaps they might. While this difficulty is not finally resolved I have suggested that the birds cannot get into breeding condition any earlier. However, the birds could continue to breed into June and July and rear a second brood at this time. Nevertheless it is clear why, in oak woodland, they do not do this. The late broods are very unsuccessful in most years and second broods would be later and more unsuccessful still. It seems highly probable that the chance of successfully raising a small number of young is not only slight, but is outweighed by the dangers to the adults. The young in the late broods are hungry and noisy and many of the broods are found by predators, and it will be remembered that of the broods taken 20 per cent had the female parent taken also. Probably, on average, those birds which do not have a second brood produce more young because more of them live to breed the next year. Thus second broods are not productive.

It is difficult to believe that the tits are laying only enough eggs to cover their losses. Adult losses are seldom more than 65 per cent, yet much larger numbers of young are produced than are necessary to replace them. In many of the years of high density a large proportion of these young probably perish soon after leaving the nest. Since these have presumably depleted the food for the others, this is not an efficient way of keeping the numbers down to a level where the food supply is conserved. For example it is estimated that, in 1961 in Wytham, 400 pairs may have raised about seven young each, only one of which was alive in the autumn; hence some 2000–2500 young great tits died between June and September.

Proponents of inter-group selection also suggest that birds limit their reproductive rate through many individuals not breeding

in the first year of life or by their being excluded (by the other birds) from obtaining a breeding site. The first of these cannot occur in the great tit since they breed when they are one year old. Also there is no evidence to suggest that there is normally a non-breeding population.

Further, it is clear that, unless the birds have very varied territorial requirements from year to year, territory size has not imposed a limit to the breeding density in most years. Since the number of pairs in Marley in 1961 was eighty-six then there must surely have been room for more pairs in all the other years when there were never more than fifty-one pairs. This suggestion is strongly supported by the fact that Hinde, working in Marley, noted that certain, apparently suitable, parts of the wood were unoccupied. Thus even if the tits were limited by territory size in 1961 (and there is no evidence that this was so) it is unlikely that they were in any other years.

Thus the evidence seems to support the theory of inter-individual selection. The birds are reproducing as rapidly as they can, and they are apparently being limited by the food resources, presumably through competition, at least at times, both in the summer and in the winter.

C. M. Perrins 247

20 R. Revelle

Population

R. Revelle, 'Population', *Science Journal*, vol. 3, 1967, no. 10, pp. 113–18.

In some sense population forecasts are the most basic of any that can be made. Nearly all the other forecasts [of world changes] will depend ultimately on the size of the human population – if only because the need for a technological advance is reflected by the number of people who require it. But however fundamental the number of people living in various regions of the world may be as a parameter, it is not an easy one to assess at the present time or to extrapolate into the future.

Like many other phenomena, changes in the size of human populations tend to follow an exponential or compound-interest law over any short period. But over longer periods populations seem to grow in more or less sudden steps, from one stage of quasi-equilibrium to another. For example, consider the two major demographic revolutions in man's experience: the population growth that accompanied the development of agriculture, and the one that began in the seventeenth century in northern Europe and has since spread throughout the world.

Man has existed in substantially his present biological form for perhaps a million years yet, during all but the last 1 per cent of that time, the number of people was never more than a few million, not much greater than the number of lions. We know this was so from the estimated density of population among peoples who were still living in a stone-age culture when they were observed or studied by Europeans – certain American Indians, African tribes, Australian bushmen, Papuans in New Guinea and others. Over most of the first million years, birth-rates and death-rates of the world's population must have been, on the average, almost exactly equal, probably somewhere between 40 and 60 per thousand people per year. The mean rate of

natural increase, taken over any set of millennia, could not have been more than 0·05 per thousand per year, corresponding to a doubling time of 20 000 years or more. Of course, the books were never so nicely balanced. In every particular region there may have been wild fluctuations from century to century, and even from generation to generation, in death rates and population size, followed in many cases by parallel fluctuations in birth-rates. But the long-term average of population size against time must have been almost a level line.

Agriculture was invented 6000—9000 years ago. Its development over the next few millennia radically changed the human condition, and destroyed the previous equilibrium between birth-rates and death-rates. In the Fertile Crescent from the Nile to the Tigris–Euphrates, in China, India, southern Europe, Middle America and Peru, human numbers may have increased a hundred fold in one or two thousand years, until a new quasi-balance between births and deaths was attained. It has been estimated that, by the time of Christ, the world population was three hundred million, even though agriculture had barely begun, or had not even been started, over large areas.

Between the beginning of our era and the early part of the seventeenth century, the population slowly increased, to somewhere around six hundred million. Then it began to rise rapidly again and at an ever-accelerating pace. Whereas the average annual rate of increase from AD 1 to 1600 had been less than 0·5 per thousand, it rose to 4 per thousand from 1750 to 1800, to 8 per thousand from 1900 to 1950, and is now probably nearly 20. The number of human beings has increased more than five fold in three and a half centuries. At the present rate it will double again by the end of our present century.

This remarkable change was brought about largely by a decline in death-rates, although fertility did rise above previous levels at some times and places. Between 1700 and 1900 the overall rate of growth was two to three times as great in Europe and areas of European settlement as in the rest of the world. Today, the situation has become almost exactly reversed. The underdeveloped countries are now growing more than twice as fast as the developed ones. At the beginning, infant and child mortalities everywhere were extremely high. In the seventeenth

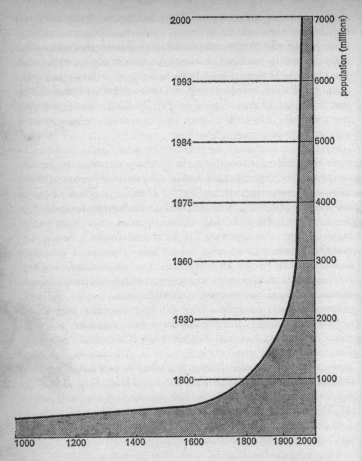

Figure 1 The population of the world has increased enormously in the past three hundred years. By 1800 – about a million years since his beginnings – man had built up a population of one thousand million. Since then, however, the periods between successive increases of a thousand million has been progressively shortened. For the population to increase from six to seven thousand million could only take seven years (1993–2000). Such explosive growth can be attributed at least in part to improved nutrition and standards of hygiene in the developing countries

century, two-thirds of the children of English monarchs, from James I to Anne, died before they were twenty-one years old. Throughout the eighteenth century, only about half the children born in Brittany – one of the 'underdeveloped' provinces of France – lived past the age of ten; elsewhere in France, for example in the south-west and in Normandy, two-thirds of the children reached their tenth birthday. In India, during the first decade of the twentieth century, the expectation of life at birth was a little over twenty years, and perhaps half the children died.

Today, in all developed countries, between 97 and 99 per cent of the children survive the first year of life, and almost all these survivors live to adulthood. In India, the expectation of life at birth now exceeds forty years, and probably at least three-quarters of the children grow up.

Even during the early stages of the modern epoch of population growth, an effective 'social' mechanism for controlling fertility existed in Western Europe – the custom of relatively late marriage for women or even life-long spinsterhood. This mechanism began to be more widely used as individuals became concerned about having too many children. The age of marriage probably rose throughout the seventeenth and eighteenth centuries; for example in 1626, in Amsterdam, 61 per cent of brides were under twenty-five years of age, but only 35 per cent a century and a half later. By 1850, the average age at first marriage in the Netherlands and Belgium was between twenty-eight and twenty-nine years, and about the same for Switzerland. From 15 to 20 per cent of women were still single at age fifty.

In several parts of Europe, a certain amount of deliberate control of fertility was apparently practised by married couples in the seventeenth and eighteenth centuries, but birth-control began to be more and more widely used and effective during the nineteenth century, first in France and the United States and later in other countries of Europe. It was clearly regarded as an improvement over delayed marriage; as fertility among married couples declined, the age at first marriage also went down; it is now less than twenty-three years in France and Belgium. In Eastern Europe, which had never adopted the Western pattern of late marriage and spinsterhood, rigorous fertility control with-

in marriage was evident in some provinces at least as early as the 1890s.

Low birth-rates and low death-rates, lower than those that prevailed anywhere at any time more than a century ago, characterize all developed countries – communist, capitalist, Catholic, Orthodox, Protestant or Buddhist. Conversely, birth-rates that were widespread two centuries ago prevail in the less-developed countries today, while death-rates are about the same as those in the developed countries forty to sixty years ago. The tragic question of our time is when and how this time gap will disappear – by a reversion to higher death-rates or an advance to lower birth-rates. The most far reaching practical task of students of human populations is to help define the biological, psychological, social and economic changes which will lead to a sufficient lowering of birth-rates in the less-developed countries.

As with any other science, the demographer's problem is to frame hypotheses that can be tested by prediction. His difficulty is similar to that of the forester or the climatologist; the events he is concerned with take a long time to happen; within one lifetime he cannot test many hypotheses by forecasting the future. In principle the difficulty could be overcome by testing hypotheses against past events – what geophysicists call 'hindcasting'.

But accurate data from the past exist over only a comparatively short time span, and are far from comprehensive. Although the Christian era began with a population count, and a few censuses were taken in Europe and North America toward the end of the eighteenth century, less than 20 per cent of the world's peoples were even moderately well counted before the middle of the nineteenth century. Even today, only about 70 per cent of human beings have been included in a census.

In this article I intend to discuss two population forecasts which have been made recently – one by the US Census Bureau and one by the Population Division of the United Nations. Both illustrate the difficulties of forecasts and the value of the information they provide. First, however, I should say something of the pitfalls of population forecasting in general. To construct a population projection for a country or region, one should know the number of people at a particular point in time, their

distribution by age and sex, their fertility, mortality and migration, and be able to make an educated guess about changes in these last three quantities during the time interval being spanned. But for large parts of the world the demographer must make do with much less information than this. Uncertainties about present size of population and rate of growth, and about future trends of mortality are three important sources of error in forecasting, and I shall discuss these in the next section. Future change in fertility is even more important, but because fertility is the more or less controllable variable in an otherwise uncontrollable situation I shall leave discussion of it till the end of this article.

The estimate of about 3300 million for the world's population as of the middle of 1965, used in both the projections I shall cite, is based on inadequate data for a number of countries. There are no recent censuses for many Asian, African and Latin American countries. Uncertainty about the size of China's population contributes a possible error of more than a hundred million. Pakistan's population, according to one estimate, was undercounted by nearly eight million (7·6 per cent of the total population) in the census of 1961. If the most probable value of about 3 per cent per year for the growth rate since 1961 is accepted, the estimate of the present population is understated by another 5 per cent. Supposing India's population to be underestimated in somewhat the same proportions as that of Pakistan, the error is more than fifty million. Taking account of the possibility of errors of a similar magnitude for many other less-developed countries, the paucity of data for others, and estimates of undercounts of 2·5 to 3 per cent for some developed countries such as the United States, the world population of 1965 may be underestimated by more than two hundred million.

The most serious problem arises from lack of satisfactory data for mainland China, which probably has nearly a quarter of the world's population. The only census of modern times was made in 1953, and gave a total of 583 million persons. There were no accurate data on the sex and age distribution of the population, and levels of fertility and mortality are unknown. Moreover, judging from experience in other countries, it is likely that the census was an undercount, probably by at least 5 per cent, and

perhaps by as much as 10 or 15 per cent. The base estimates as of mid-1953 for the population of China would then be between 610 and 640, or even 670 million. On the other hand, if the published figure of 583 million is taken as the midpoint of the probable range, with uncertainty of plus or minus 10 per cent, the spread of possible population size in 1953 would be 525 to 640 million.

If all the possibilities for China's population are incorporated into alternate estimates and projections, the totals at the upper and lower limits describe an extremely wide range. Projections for 1985 have a spread of more than five hundred million, with no guarantee that all possibilities have been covered. One demographer has estimated a possible range from one thousand million to two thousand million in the year 2000, while the United Nations projections give a minimum of under nine hundred million and a maximum of over fourteen hundred million.

Of equal significance are the rates at which populations are now growing. Estimates of present growth rates for much of the world are of doubtful reliability or are simply not available. Very divergent values may be obtained by different techniques. For example, the rate of growth for Pakistan is 2·1 per cent per annum if one accepts the growth indicated by the censuses of 1951 and 1961, but about 3·2 per cent if one accepts the results of sample surveys conducted since 1962. A population increasing at 3·2 per cent annually will be 12 per cent larger at the end of ten years than one of the same initial size growing at an annual rate of 2·1 per cent.

Since the end of the Second World War, there has been a very sharp decline in mortality levels in many developing countries. This is usually ascribed to large-scale public-health measures, together with the widespread use of antibiotics and insecticides, although improved foods and better food distribution, particularly for children, have undoubtedly played a part. The pace of mortality decline has probably slackened in the past two or three years, at least in several developing countries. Major improvements in standards of living, and more equitable income distribution, may be necessary for a future marked reduction of mortality.

Nearly all countries are committed, in one degree or another,

to further reductions. There is no certainty, however, that this commitment can or will be translated into markedly lower death-rates. Population is already pressing hard on the food supply of a number of nations. Except for shipments of grain from the United States, Canada and a few other countries in 1966, death-rates in India might have risen sharply.

As food requirements increase, the chances that a country will be able to forestall a major famine by importing food from abroad may diminish. The absence of large reserves, which characterizes the present situation in developed countries with surplus producing capacity, lessens the likelihood that a subsistence crisis in less-developed countries, produced by drought or flood, can be met. Even if production in the developed countries could be increased to meet the need, their electorates might be unwilling to accept the economic burden of filling the food deficits of the poor nations.

Two series of projections of the population of India, Pakistan, Brazil and the world in 1985 have recently been made by the United States Census Bureau for President Johnson's Science Advisory Committee. The high series were based on the assumption that fertility will remain at present levels, while fertility was assumed to decline in the low series. It was assumed that mortality would decline at the same rate in both.

For the assumed growth rate of the world population in 1965, the United Nations estimate of 1·8 per cent per year for the rate of growth from 1960 to 1964 was used, but other sources were taken for estimating present growth rates in Brazil, Pakistan and India. The estimate for Brazil was about 2·9 per cent, for India about 2·6 per cent, and for Pakistan 3·2 per cent.

The level of mortality for the world in 1965 was taken as 16 per thousand population; the birth-rate would then be 34 per thousand, about equal to that of Taiwan. A set of female age-specific fertility rates was obtained by slightly modifying a schedule of known rates for Taiwan in 1960. For the high projection this was assumed to remain constant but for the low one it was postulated that fertility rates would decline to 90 per cent of the 1965 values by 1970, to 80 per cent by 1980 and to 70 per cent by 1985.

For both projections for Pakistan the expectation of life at birth was assumed to increase about 3·3 years per decade and the levels of fertility for East and West Pakistan were considered separately. East Pakistan's birth-rate was assumed to be 53 per thousand in 1961 (the time of the last census) while that for West Pakistan was placed at 50. For the low series it was assumed that the intensified family-planning programme introduced in 1965 would be gradually extended throughout the population, and that its maximum effectiveness would be reached by 1972, at which time the fertility rate would have decreased 28 per cent. After 1972 the family-planning programme was assumed to expand at the same rate as the population; in other words, the proportion of the population practising family-planning techniques would remain essentially constant.

The Indian high projection was based on estimates by the Institute of Applied Manpower Research in New Delhi. The average birth-rate during 1961–5 was assumed to be 41 per thousand and this was taken as the 1965 birth-rate. The expectation of life at birth was assumed to increase about seven years per decade. For the low projection, the percentage decline in fertility assumed for Pakistan was adopted.

For Brazil, the United Nations Economic Commission for Latin America estimated the birth-rate during the period 1959–61 to be in the range of 40 to 43 per thousand. A birth-rate of 41 for 1965 was therefore assumed to establish the level of (constant) fertility for the high projection. For the low projection, a modified version of the fertility model used for Pakistan was employed.

The high projection shows an increase in the population of the world from 3300 million in 1965 to 5030 million by 1985, or by 52 per cent. The low projection gives a population of 4645 million by 1985, an increase of about 40 per cent above the 1965 figure. These correspond to average annual growth rates over the twenty-year period of 2·1 per cent and 1·7 per cent respectively. The difference between the high and low figure is only 385 million persons, about the same as the likely range of uncertainty in the projected population of mainland China in 1985. However, the difference would rapidly widen in later decades if the 1985 differences in rates of increase were to persist. By the year 2000, for example, an annual rate of increase of 2·4 per cent,

starting with a population of 5000 million in 1985, would give a world population of 7150 million persons, whereas an annual rate of 1·7 per cent and a 1985 population of 4650 million would result in a world population of 6000 million by the year 2000.

The proportional difference in population size between the high and low projections for 1985 was greater in India and Pakistan than in the world as a whole; the high value for both countries combined being about 1100 million, 12 per cent greater than the low one of 980 million. By the end of the century, if the 1985 rates of increase were to continue, the population of the subcontinent would be nearly 25 per cent greater under the high projection than the low. In the year 2000, the population of over 1800 million persons under the high projection would be three times the present population of the region, more than half the 1965 population of our entire planet, and perhaps considerably higher than the population of mainland China at the end of this century.

In the developed countries, with their low birth-rates, there are between two and three adults (persons twenty years of age or older) for each child under fifteen. In the less-developed countries, the number of adults is about equal to the number of children.

During the next two decades, the proportions of children in the developed countries will probably diminish, and their absolute numbers will increase by only around 15 per cent. In contrast, if fertility remains at present levels in the less-developed countries, the proportion of children will rise, and their absolute numbers will be about double. On the other hand, if fertility can be reduced by the amounts we have assumed, the ratios of children to adults in Brazil, India and Pakistan will diminish by 12–16 per cent below present levels, and by about 20 per cent below the ratios for constant fertility. The increase in numbers of children will be only about one-half to two-thirds the increase that will occur if fertility remains constant.

These possible differences in numbers of children have serious implications for education. In Pakistan, for example, under the high projection the numbers of children of primary-school age – five to fourteen years – will increase by 118 per cent in twenty years or 4 per cent per year. This means that with an annual

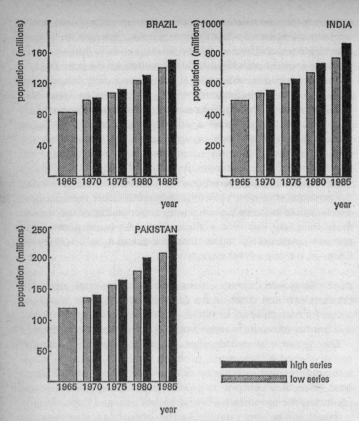

Figure 2 Projections of the populations of Brazil, India and Pakistan up to 1985. The data were prepared by the US Census Bureau for President Johnson's Science Advisory Committee. The high series was based on the assumption that fertility will remain at present levels, while fertility was assumed to decline in the low series. Mortality was assumed at the same rate in both. Extrapolation from the high series for India and Pakistan suggests that by the year 2000 the population of the subcontinent could be over 1800 million – three times the present population of the region, more than half the 1965 population of the world and perhaps considerably higher than that of mainland China at the end of the century

economic growth rate of, say, 5 per cent and a constant fraction of national income going into education, the educational expenditures per child of primary-school age can be increased only 1 per cent per year, 22 per cent in twenty years. Since less than half the children are now in school, the proportion receiving a primary education in 1985 will still be less than 60 per cent. The fraction of national income going into education will need to be nearly doubled to ensure even a primary education at the present minimal level for all the children, let alone to improve the quality or lengthen the period of schooling.

Today's poorly educated children in the less-developed countries will be tomorrow's job seekers. According to our estimates, India's population in the age group fifteen to thirty-four will increase over the next two decades by 107 – 116 million, Pakistan's by about 41 million and Brazil's by 22 million, a total of 170 – 180 million. This means new jobs must be created for about one hundred million young persons entering the labour force for the first time. The total size of the labour force in these three countries will grow by roughly two hundred million, or 70 per cent. Nearly all this increase will come from persons who are already born; consequently, the magnitude of the employment problem in 1985 will not be much affected by changes in fertility levels during the next twenty years. Assuming a capital investment of $1000 is required on the average to create a new job, the total investment needed for the three countries by 1985 would be at least $200 000 million to maintain even the present very low *per capita* incomes.

The Population Division of the United Nations has made high, low and medium estimates of the world population for each decade up to the year 2000. To calculate these projections, the world was divided into twenty-four regions, in each of which separate estimates of probable future mortality and fertility were made. In the developed countries, the levels of birth- and death-rates were expected to undergo very little change; the decennial rate of natural increase, which was 11 per cent in the 1960s, was assumed to decrease slightly to 9 per cent in the 1990s. In the developing regions, considerable changes were expected in the main components of population growth.

To project mortality levels, it was assumed that average life expectancy at birth of a population would rise annually by a half year until it attained fifty-five years; between fifty-five and sixty-five years, there would be a slightly higher gain each year, followed by a slow-down after sixty-five until the gain became negligible when life expectancy had risen above seventy-four years. On this basis, death-rates in the poor countries, which currently average twice as high as in the developed nations, were expected to drop from 18 per thousand at the beginning of the present decade to 10 per thousand in the 1990s, by which time they would have attained the same average as those of the developed countries.

For many poor countries, fertility has not yet started to decline. In these it was assumed that once a decline begins, thirty to forty-five years are needed to drop to half the original level. As to the date of onset, several assumptions were made for each region. These different dates, combined with different assumptions concerning the duration of decline, give rise to the high, medium and low projections. The medium projection gives an overall decrease of one-quarter in the average birth-rate of the poor countries, from 40 per thousand in the 1960s to 29 in the 1990s, that is, from a little over twice the present birth-rate of the developed regions to a little over 1·5 times this level.

In the medium projection, the world's population, totalling 3281 million in 1965, is estimated to rise by 87 per cent to 6130 million by the year 2000. The high and low projections, which the UN regards as equally plausible, might raise the world population by as much as 113 per cent at the century's end, to 6990 million, or by as little as 66 per cent, to 5450 million. (In the past, the UN high projections have turned out to be closer to reality than its medium projections.) If fertility remained at present levels in every region, the world total would be 7520 million by 2000, with ever-accelerating growth, but this was regarded as unlikely. The increase during the remainder of the century would then be 129 per cent of the 1965 world total, 53 per cent in the developed regions and 164 per cent in the developing ones. Some demographers have taken a very much more optimistic view about possible future declines in fertility, and have projected a world total at the end of the century of between 4200 and 5000 million

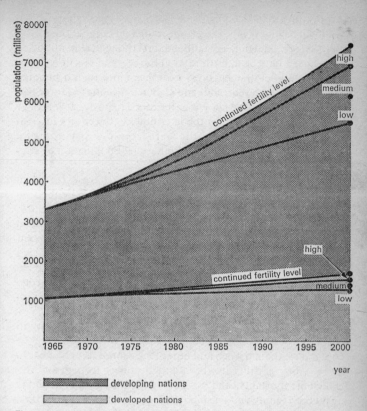

Figure 3 Projections of population for the world (total shaded area), developed (light shade) and developing nations (dark shade); these are based on figures produced by the Population Division of the United Nations and the US Census Bureau. The data for the year 2000 show the four variants – low, medium, high and continued fertility level – postulated by the United Nations. At the present time one-third of the world's population lives at the high standard of the present developed nations; by the end of the century only one-quarter of the population will have this advantage

people. This demonstrates clearly the importance of fertility control and it is now time I dealt with the problems this involves in more detail.

Fertility rates in most human societies are well below the potential of human fecundity; the direct constraints on population growth are social ones and not, as Malthus at first thought, starvation and disease. As I have said, birth-rates in Europe and North America began to fall below their previous levels more than a hundred years ago. In the seventeenth and eighteenth centuries, there had been a growing tendency in Western Europe to postpone marriage and to increase celibacy. In the later nineteenth century married couples began to resort to abortion on a large scale and to practice contraception, in addition to the limitation of population by emigration.

In recent years, rapid declines in fertility have occurred in Eastern Europe and Japan. Fertility rates went down by about 40 per cent in Hungary between 1954 and 1962, by 36 per cent in Rumania between 1955 and 1962, and by 30 per cent in Poland between 1955 and 1964. Japan experienced a fall of 44 per cent between 1950 and 1962. The conditions in these nations differed markedly from those that prevail today in the underdeveloped world. Literacy rates were high, many couples were already deliberately controlling their fertility and they had strong motivations to limit family size.

Yet there are good reasons to believe that human fertility in most of the less-developed countries will decrease during the next few decades, though when and how much cannot be predicted. In the past, high fertility has been an adjustment to high and unpredictable mortality, and has been well fitted to a community life built on family and kinship ties. Only one son may have been needed for ritual or economic purposes, but it was common to desire at least two for insurance against the death or incapacity of one. Families had to average four children to obtain two sons. Clan and tribal kinship systems enhanced the value placed on children, because they contributed to the power or status of the group. Today, as can be seen from surveys of individual husbands and wives, many couples in the less-developed countries want no more children than they already have, and would like to know better ways of preventing further births. At the same time, these surveys show that the desired numbers of children are much higher than in the developed countries. The average is four children per couple, which, with present and prospective levels of mortality,

would result in a doubling of population every thirty to thi.
years.

Most people in developed countries want from two to four
children; in the less-developed countries they want from three to
five. In the developed countries, actual family size is slightly less
than desired family size, while the reverse is true in less-developed
ones. In the poor countries, favourable attitudes to family limitation are much more common among couples who already have
four or more living children. Availability of contraceptive devices
may not be very important until the desired number of living
children is secure. Low infant and child mortality, and public
awareness that mortality is low, may be one of the necessary
preconditions for reducing fertility.

In developed countries, infant mortalities are nearly always less
than 50 per thousand live births. But in the less-developed
countries, with very few exceptions, 50 to more than 150 children
out of every thousand die before the age of one. In a family of
four or five children the probability that all will grow up is often
less than 50 per cent.

When average infant and child mortality is high, the variance
is also high: the chances in an individual family that several
children will die are frighteningly large. Because parents desire a
high degree of assurance that one of their sons will grow up to be
a man, they are willing to assume the burden of having too many
children in order to gain this assurance. We are faced with the
apparent paradox that a reduction in mortality may reduce rather
than raise the rate of population growth. The population problem
of our time has been created by lowering death-rates, yet an
essential element in curing it may be to lower infant and child
death-rates further still, perhaps down to the levels existing in the
Western world. If this is so, then the quickest possible increase of
food supplies, both in quantity and quality, is of the utmost
urgency for the long term as well as the short, because poor
nutrition in the poor countries strikes fiercely at the children. In
the developed countries, most human beings live to middle age,
and cardiovascular diseases and neoplasms are overwhelmingly
the leading causes of death. In developing countries, the main
killers are diseases of childhood resulting from a combination of
infection and malnutrition.

R. Revelle 263

all proportion of people in the less-developed coun- even moderately good knowledge of modern methods of anning; the poor and the uneducated need to learn what -to-do and the educated already know – that there are a number of safe , reliable and simple ways of limiting one's family. Knowledge of contraceptive methods is much rarer than the desire not to have more children.

The governments of developing countries are now adopting population-control policies at a rate and in a climate of world approval unimaginable even a few years ago. Among the nations that have officially decided to foster family planning are India, Pakistan, mainland China, South Korea, Ceylon, Singapore, Hong Kong, Malaysia, Turkey, Egypt, Tunisia, Morocco and Honduras. Taiwan has no formal policy, but the Government has given full cooperation to an island-wide programme that has already reached a substantial part of the population. In many countries, at least the beginning of governmental interest is visible. These include the Philippines, Thailand, Nepal, Afghanistan, Iran, Kenya, Mauritius, Chile, Colombia, Peru and Venezuela.

The role of governments in reducing fertility is to exhort, in-form and provide: decisions and actions must be taken by indi-vidual couples acting in accordance with their perceived interests. Even so, the governmental task is large and difficult, requiring a high degree of organization, adequate financial and logistic sup-port, great flexibility in meeting changing conditions, and con-tinuing objective evaluation of results.

Although there have been high hopes for family-planning pro-grammes using intra-uterine devices, and birth-rates appear to have gone down in some countries, notably Korea and Taiwan, the part played by these birth-control programmes is not clear. Many of the devices may have been used by older women who were already controlling their fertility in other ways. Postponing the age at which women have their first child, and longer spacing between pregnancies, are now being recognized as important. In Japan, postponement of marriage has been one of the key factors (together with abortion) in the post-war fertility decline. In the Republic of Korea, an increase in the mean age at marriage of women from nineteen years in 1935 to twenty-four years in 1960

has apparently accounted for a 16 per cent drop in total .
rates.

We need to find, to learn how to bring about and how to help people recognize the changes in living conditions in the less-developed countries that will lower the benefits to individual families of having more than two or three children, and increase their costs. At the same time, research is needed to develop methods of fertility control that are easier to introduce than the present oral contraceptives and mechanical devices. Much of this research can be done by the developed nations.

Moore

Communication in Insects

B. P. Moore, 'Chemical communication in insects', *Science Journal*, vol. 3, 1967, no. 9, pp. 44–9.

The amazing ability of some male moths to detect females of their own species at distances of several thousand metres has long intrigued naturalists. At the turn of the century the great French observer Jean Henry Fabre was able to show that the attractiveness of the virgin females for their prospective mates was chemically based. But the amounts of natural lures involved were so minute that progress towards chemical identification proved impossible for over fifty years. However, the recent advent of sophisticated physicochemical techniques for dealing with trace components has given the research new impetus. Results over the past ten years have been spectacular: several insect sex attractants have been chemically identified and synthesized in pure form, and behavioural experiments have established the existence of many others in a wide range of insects. Moreover, chemical mediators of various kinds have been found to play an absolutely crucial role in the biology of social insects. On present trends it may fairly be claimed that chemical messenger substances – or pheromones – will prove to be important factors in the life histories of the majority of insects and will be found to have contributed enormously to their evolutionary success in a continually changing world.

Pheromones can be classified according to the effect they produce upon the recipient animal. The 'releaser' pheromones give rise to a more or less immediate, though temporary, change in behaviour, and they are presumed to act via the recipient's receptors and central nervous system. These are mostly relatively simple, volatile substances that transmit their message via the olfactory system, and they form the basis of most short-term chemical signals

in both social and non-social insects. They include the sex attractants and sex activants, trail-laying scents, alarm substances and marker scents, many of which have engaged the attentions of chemists in recent years. A few have so far been identified and their structures prove rather diverse, although some overall resemblance is discernible between analogous pheromones from related species.

'Primer' pheromones, on the other hand, produce little or no obvious behavioural response but serve instead to trigger or govern a train of physiological changes in the recipient – changes which in turn may ultimately lead to profound changes in development and the associated behavioural repertoire. Most primer pheromones appear to be transmitted by mouth, but it is as yet uncertain whether they act directly upon the endocrine systems or are mediated by taste receptors and the central nervous system. Primer pheromones are restricted to social insects, where they regulate the production of the different castes, but their indirect mode of action makes them difficult to study and progress with chemical identification has been slow. Only one, the 'queen substance' of the honey-bee, has been identified to date.

Most present-day insects lead an essentially independent existence and the need for collaborative effort between them is usually restricted to conjugation of the sexes, although general aggregation for such purposes as hibernation has evidently proved advantageous with species showing 'warning' colours, where a massed effect is doubtless more efficient as a deterrent to predators. These activities are often mediated by pheromones and the problem is to account for the elaboration of such substances, and in particular the highly active and specific sex attractants, during the course of insect evolution.

Early primitive insects were probably omnivorous or ate rotting materials and were cryptic or secretive in their habits; they showed little metamorphosis but probably continued to moult at intervals throughout their lives. Moreover, the poor water-retaining properties of their cuticles would have tied them to restricted and sheltered habitats, such as beneath stones, litter or under loose bark. Mating was probably indirect, involving deposition by the male and subsequent taking up by the female of a

preformed 'packet' of sperm or spermatophore, as in present-day silverfish of the order Thysanura. The sense of smell of these early insects would not have been developed. Food would have been located by random searching and the sexes would have met by chance.

With the development of more specialized cuticles and, later, the evolution of wings, it is likely that insects were released from their restricted environment: they became more mobile and better able to adapt themselves to a wide range of discrete habitats. Feeding habits would also have become more specific and these, in turn, must have entailed an increase in powers of food detection and discrimination. In many cases, these adaptations would have involved at least an element of olfactory, or scent, perception and, as a corollary, the meeting of the sexes would have been made easier by their aggregation in restricted habitats or on the food material. Subsequently, evolution would have proceeded towards the exploitation of rich but short-lived food supplies, such as fungi or carrion, with continual sharpening of the insects' olfactory powers.

Now insects with highly developed olfactory systems would have been pre-adapted, as it were, towards chemical means of communication and, as a further stage in the evolution of attractant pheromones, one might envisage the situation where the activities of one sex would render the food even more attractive to the other. One sex (usually though not invariably the male) would be attracted by one or more metabolites of the other and evolution, through a series of small steps, would ultimately lead to combination of the several signal components into a single highly specific chemical structure. At this stage, the species could well have become independent of the food lure and further mutual sharpening of both scent and receptor mechanisms could lead to the specifically tailored sex-attractant molecules characteristic of many Lepidoptera (butterflies and moths) and Coleoptera (beetles) today.

It is interesting to examine the pheromones of various groups of modern insects in the light of the postulated mode of evolution and, first, I shall take the silverfish and their allies (orders Thysanura and Archaeognatha) which are undoubtedly the most

primitive of living insects. Most of the species, apart from a few domesticated forms, have departed but little in overall development and biology from the ancestral condition, and chemical communication is not well marked. There is, nevertheless, some behavioural evidence to suggest that pheromones may be responsible for synchronizing the moulting cycles in mature prospective pairs, so that more frequent and effective mating may take place.

Coleoptera, in general, show remarkably close adaptation to restricted habitats and food supplies and they afford examples of transition between feeding and non-feeding adult forms, with attendant development of attractant pheromones. Thus in certain bark beetles of the family Scolytidae, notably the pine-bark beetle (*Ips confusus*), pioneering males, after feeding upon a suitable host tree, produce a pheromone attractive to both sexes, thereby causing aggregation of breeding populations. This pheromone is apparently a blend of several related monoterpenoid (C_{10}) alcohols which, when separated, show no individual activity. The mixture is produced in the male hind-gut and it may well result from simple metabolism of one or more constituents of the host plant.

Adults of many Australian chafers are attracted to a common food source, usually *Eucalyptus* foliage, where mating takes place; such species would have little use for sex attractants and are not known to produce them. However, species with a wide choice of food plants, or species in which females are flightless, would benefit from a sex-attractant mechanism and in some species that have been examined appropriate pheromones have been detected. Chafers of the genus *Rhopaea* are a case in point and the three species investigated by Roberts and Soo Hoo show the transition nicely. None of the three feeds in the adult stage but females of all of them produce a sex attractant. Females of two species (*R. magnicornis* and *R. morbillosa*) seldom fly, but both sexes of *R. verreauxi* are attracted to snow-gum foliage, where they often mate. The alimentary canals of all three species are non-functional and show degeneration but, as might be expected, that of *R. verreauxi* is the best developed. This species is evidently nearest to the ancestral habit of meeting and mating on the food material.

B. P. Moore 269

The Lepidoptera furnish some of the best documented and most spectacular cases of long-range chemical communication, but here again we can detect, in present-day species, various apparent stages and directions of evolution. At one extreme there are the diurnal, nectar-feeding butterflies that appear to have evolved in step with flowering plants. Here, visual perception is important and acute and it mediates both flower recognition and courtship. Conversely, pheromones are poorly developed and are apparently limited to sex activants produced in the special wing-scales (androconia) of the males.

Many moths are also nectar feeders but most are twilight or nocturnal forms. With these species, olfactory responses assume an increasing importance and the visual element in food-plant recognition correspondingly declines. These are the species that are attracted in large numbers to the naturalist's well-known and highly aromatic sugaring mixtures and, as would be predicted from my general thesis, short-range sex attractants and activants are widespread amongst them.

Lastly, we come to species that do not feed at all during the adult stage and, significantly, it is here that the highest development of specific sex attractants is reached. In some cases the females have become secondarily flightless and dependence upon a sophisticated lure-and-receptor system is then complete. Males are characterized by their very feathery antennae, the seat of numerous olfactory receptors. The sensitivity of these receptors is quite remarkable: in the case of the gipsy moth *Porthetria dispar* a few hundred molecules of gyptol, the female sex lure, per cubic centimetre of air (that is, about one part in 10^{17}) are enough to elicit the male response! The degree of specificity attained is such that only males of very closely related species are attracted by a given lure; these species are kept apart in nature by seasonal and ecological differences.

The sex attractants of four moth species, the silkworm moth (*Bombyx mori*), the cabbage looper (*Trichoplusia ni*), the gipsy moth and the pink bollworm moth (*Pectinophora gossypiella*), have now been identified and synthesized; they are all unsaturated primary alcohols or their derivatives and thus show an intriguing similarity in structure. As the moths themselves are not closely related, but belong to three superfamilies (Bombycoidea, Noctu-

oidea and Gelechioidea), it is tempting to infer that this type of structure is general throughout the Lepidoptera. At any event, these sex-attractant pheromones represent the ultimate in chemical communication amongst non-social insects.

The elaboration of pheromones has also proved a key factor in the establishment of communal life among insects and it is therefore to the social insects – the termites, ants, wasps and bees – that we must turn to witness the greatest sophistication in chemical communication. The huge persistent colonies characteristic of these insects have been likened to a super-organism in which individual animals take the place of the usual cells. The development and behaviour of these individuals are regulated by pheromones much as the cells of the body are governed by the hormones. The centre of organization, the queen (or in termites, the royal pair), may be likened to the brain.

Here, too, we can detect in living insects almost every phase that might be postulated in evolution between the most rudimentary associations of individuals and advanced social communities. Moreover, it is very evident that there has been parallel evolution, in their way of life, between otherwise remotely related groups. Thus on structural and fossil evidence, the social termites may confidently be associated with the non-social cockroaches; yet the former have evolved a communal mode of living – based upon pheromones – and superficially remarkably akin to those of the social Hymenoptera (bees, wasps and ants).

Present-day locusts appear to represent an early stage in evolution of social living, for at low population densities they develop a 'solitary' phase, with a largely independent existence. But as living conditions in the desert improve, and crowding occurs, there is a switch to the 'gregarious' phase where development and behaviour are modified by frequent interactions between hoppers – and adult swarming ultimately takes place. Admittedly these changes are largely mediated by visual and tactile stimuli, but already pheromones are beginning to play an integrating role, particularly in the choice of favoured oviposition sites. These pheromones have not yet been identified.

Professor Michener of the University of Kansas has postulated that social behaviour in bees of the subfamilies Halticinae

Table 1 Some Insect Pheromones of Known Structure

Bombycol	$CH_3CH_2CH_2CH=CH-CH=CH-CH_2CH_2CH_2CH_2CH_2CH_2CH_2OH$	sex attractant of the female silkworm moth (*Bombyx mori*)
Gyptol	$CH_3CH_2CH_2CH_2CH_2CH_2CH-CH_2CH=CH-CH_2CH_2CH_2CH_2CH_2CH_2OH$ $O\cdot CO\cdot CH_3$	sex attractant of the female gipsy moth (*Porthetria dispar*)
Propylure	$CH_3CH_2CH_2$ $C=CH-CH_2CH_2CH=CH-CH_2CH_2CH_2CH_2O\cdot CO\cdot CH_3$ $CH_3CH_2CH_2$	sex attractant of the female pink bollworm moth (*Pectinophora gossypiella*)
Farnesol	$CH_3C=CH-CH_2CH_2C=CH-CH_2CH_2C=CH-CH_2OH$ $CH_3CH_3CH_3$	sex attractant of the male humblebee (*Bombus terrestris*)
Queen substance	$CH_3CO\cdot CH_2CH_2CH_2CH_2CH_2CH=CH\cdot CO\cdot OH$	caste-control pheromone and sex attractant of the queen honey-bee (*Apis mellifica*)

Citral	CH_3 $CH_3$$\backslash$ $C=CH-CH_2CH_2C=CH-CHO$ $CH_3$$/$	alarm pherome of the leaf-cutter ant (*Atta sexdens*)
Limonene	(structure of limonene)	alarm pheromone of the Australian harvester termite (*Drepanotermes rubriceps*)

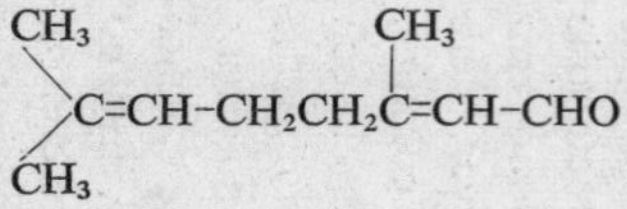

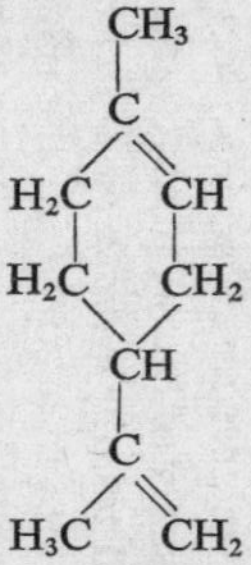

Citral

CH_3
$\quad$ CH_3
$C=CH-CH_2CH_2C=CH-CHO$
CH_3

alarm pherome of the leaf-cutter ant (*Atta sexdens*)

Limonene

alarm pheromone of the Australian harvester termite (*Drepanotermes rubriceps*)

pinae has developed through a series of steps in association
...en adults, firstly to form semi-social groups and ultimately
...ing to true societies. He discovered, in his field investigations,
almost every phase necessary to link gregarious though independently nesting species, on the one extreme, to fully integrated
social forms on the other. The early phases in association were
presumably based mainly upon visual stimuli but pheromones
would be implicated increasingly as integration proceeded.

By contrast, social behaviour in termites, wasps and ants
appears to have evolved from primitive adult–larva associations,
with trophallaxis (exchange of nutrients) as the main motivating
force. Such associations still seem to be much more important in
these groups than in most bees and, in some lower termites,
juvenile forms still provide the worker caste.

I can scarcely overemphasize the importance of pheromones in
communal insect life, for these chemical mediators form the very
basis of controlled development and integrated effort within the
colonies. Combined with subsidiary visual, auditory and tactile
signals they truly form the 'spirit of the hive'. Primer pheromones
may well have proved the mainspring in evolution from solitary
to social living, enabling certain groups of insects to achieve a
level of adaptive success and dominance far above the general
level of their relatives.

The mode of action of primer pheromones, at the metabolic
level, is still shrouded in mystery and their subtle, long-term
activities make them difficult of access and study. However, there
is much evidence, derived from manipulated laboratory colonies,
to confirm the existence of quite complicated networks of such
pheromones and to underline their importance in maintaining the
delicate caste balance characteristic of social-insect populations
in the field. Dr Martin Lüscher of Berne, in his classical researches
on the European drywood termite *Kalotermes flavicollis*, invoked
no less than five pheromones to account for the paths of control
and replacement of reproductive individuals he observed in this
species – and further pheromones would be needed to cover the
soldier caste!

In the honey-bee, on which so much painstaking research has
been conducted, the queen is a veritable font of pheromones,

most of which are produced by her mandibular glands. One of these, the 'queen substance' has been identified as 9-keto-2-*trans*-decenoic acid and is now made synthetically; it is an ambivalent pheromone, that is to say it exhibits more than one specific function. As a primer pheromone, it inhibits the development of ovaries in worker bees and the construction, by workers, of new queen cells. If present throughout the hive in sufficient concentration it effectively blocks the rearing of supernumerary queens, but whenever the concentration falls below the threshold level, either as part of the swarming cycle or as a result of senescence in the reigning queen, new queens are soon produced. Queen substance is very attractive to worker bees, who seek it avidly from each other and from the queen, thus ensuring its uniform distribution throughout the hive; it also functions in a releaser role as a sex attractant for drones on the mating flight.

Our knowledge of releaser pheromones in social insects is much more extensive, for the direct behavioural responses produced by these short-term mediators usually afford a basis for a suitable biological test. The detection of such responses is relatively straightforward and the pheromones themselves are often present in readily workable amounts. Alarm pheromones are a case in point; they are usually simple volatile compounds that function at rather high concentrations, for their messages need to be sudden but short lived. A number have been chemically identified: iso-amyl acetate is a component of the sting pheromone of the honey-bee; and limonene is the alarm substance of the Australian harvester termite (*Drepanotermes rubriceps*); 2-heptanone, methyl heptenone, citral and citronellal are responsible for alerting different species of ants.

Efficient direction finding and attraction to promising feeding sites are essential to social insects that forage, and these activities, too, are mediated to a greater or lesser degree by releaser pheromones. In the honey-bee, direction finding is based upon perception of the plane of polarized sunlight and transfer of information concerning locations of distant food sources involves the famous 'waggle dance'. But worker bees use a pheromone from their Nassanoff gland (on the upper side of the abdomen) to attract others to newly discovered food sources. The chemicals involved are chiefly geraniol, citral and nerolic acid.

All species of termites so far examined prove to possess an abdominal scent gland that is used in laying odour trails. At first, this seemed surprising for many lower termites do not forage but live within the rotten timber that forms their food. However, Alistair Stuart showed that in such species scent trails serve to recruit soldiers and workers to breaches in the nest walls, and this he believes to be the ancestral habit. Higher termites have secondarily adapted this trail-laying mechanism to foraging purposes. I have recently isolated the scent-trail pheromone common to several Australian species of *Nasutitermes*; it has the constitution of a diterpenoid (C_{20}) hydrocarbon and it gives rise to surprisingly persistent trails, with a life of several days under laboratory conditions. It is present in the termites at the level of about one part in four million and is effective in concentrations of from 10^{-5} to 10^{-8} grammes per millilitre of inert solvent.

In studies with laboratory colonies of ants, Professor E. O. Wilson has demonstrated the existence of numerous other pheromones, mediating such responses as colony recognition, trophallaxis and disposal of the dead, and similar mechanisms will doubtless be discovered in other groups of social insects. Indeed, every new investigation in this area serves to emphasize the maxim expressed earlier in this article, that chemical communication is the mainstay of social insect life.

We are witnessing, in pheromone research, the growth of a new and exciting discipline in biology, where the combined efforts of chemists, physiologists and entomologists promise a lead to much better understanding of the insect and a new approach to the problems of pest control.

22 H. C. Bennet-Clark and A. W. Ewing

The Love Song of the Fruit Fly

H. C. Bennet-Clark and A. W. Ewing, 'The love song of the fruit fly', *New Scientist*, 26 October 1967, pp. 230–32.

Much of our present-day knowledge of heredity has been gained through the study of a little fly, *Drosophila melanogaster*, variously called a fruit fly, a vinegar fly, a pomace fly and a garbage fly. It is a harmless creature about two millimetres in length that breeds rapidly under laboratory conditions. It has a generation time of under two weeks and one fertilized female can produce up to two hundred offspring. Many easily identifiable mutant characters such as differences in eye colour have been described, which makes *Drosophila* an ideal organism to demonstrate Mendelian segregation. Also, these flies have in their salivary glands giant chromosomes containing dark bands which can be considered as indicating the positions of genes. Translocations, inversions and deletions within the genetic material can easily be seen in these giant chromosomes.

Drosophila is usually cultured on a corn meal–molasses–yeast mixture in half-pint milk bottles which are plugged with cotton wool. A pair of flies can be placed in a bottle, allowed to mate and lay eggs. When this was done, originally, with flies caught in the wild a few pairs refused to breed and it was quickly discovered that there were many different species of *Drosophila* which were superficially very similar.

In America Sturtevant described twenty-three new species in a single paper in 1916, and since then perhaps some two thousand species have been recorded, almost one thousand of them from the island of Hawaii alone, where evolution in the genus *Drosophila* appears to have run riot. Many species resemble one another so closely that they can only be separated by careful examination under a high-power microscope or even, in some

cases, by dissection and comparison of internal features. Such closely related species are termed sibling species.

If this is the only way they can be separated how do the flies distinguish potential mates of their own species? It has been found that they do this partially by scent or taste but also we believe by the nature of their songs.

If you watch a pair of *Drosophila melanogaster* you will see that before copulating the male performs a courtship dance. During this he extends one wing and runs after the female vibrating this wing. If she has not previously mated and if she is mature, she will stop running and allow the male to lick her genitalia with his proboscis and then to mount her and copulate. [. . .] If the two flies are of different species the male will probably court but the female will jump or fly, or kick, or flick her wings at the male.

Thus the courtship is one where the male will display to a female and she determines whether or not he is a suitable mate from his display. The American zoologist Spieth, has described the courtship displays in about forty species of *Drosophila* and he found that many different species had seemingly identical displays. It was unlikely therefore that the females could discriminate on the basis of visual stimuli. This was supported by the finding that the majority of species mate quite successfully, and without hybridizing, in the dark.

In 1964 one of us found that courtship success was related to the wing area of the male. A male with amputed wings mated more slowly and a specially selected large-winged stock provided males that mated faster than normal flies. About this time Shorey found that male *Drosophila melanogaster* made a noise during courtship consisting of a series of pulses of regular duration and interval. We became interested in this as it seemed likely that the pulses were produced by the wing display and were a code that identified the male to the female.

The first part of our work was to build apparatus to record the courtship sounds. We quickly found that the sound was very low indeed – the wing is only about 2 mm long and beats through about 0·5 mm. A crystal microphone was sensitive enough to pick up the sound if the flies were placed directly on the microphone diaphragm but that 200 mm of plaster board and glass wool sandwich all round barely gave enough insulation.

The 'song' of *Drosophila melanogaster* is a series of nearly pure single sinusoidal cycles of sound (Figure 1). At 25 °C, the single cycle has a frequency of about 330 Hz and is repeated twenty-nine times a second. Its sibling species, *Drosophila simulans* produces the same sound pulse twenty times a second. Here, then, were two closely related species with a similar display differing only in pulse repetition rate (Figure 1).

At this stage there was only circumstantial evidence that this sound was relevant in courtship, so we built an electronic circuit

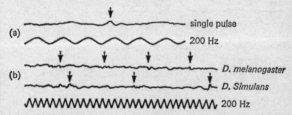

Figure 1 Courtship sounds of two species of *Drosophila*. In (a), a single sound pulse is shown, in (b) *D. melanogaster* and *D. simulans* show pulses produced at different rates

that produced an electrical analogue of the courtship sounds and used this to drive a loudspeaker.

The artificial courtship sound was played to groups of normal females and wingless males and these were induced to mate more rapidly than without the sound. When an air current was blown over them as well, the flies mated even more rapidly than the normal winged males. So it appears that the winged display provides tonic and phasic stimuli and, with the right sound, a female will accept an otherwise unattractive male.

Since then we have examined other species which are closely related to *D. melanogaster* and *D. simulans*. Most of these produce trains of single sine waves of sound repeated, as for example in *D. bipectinata*, as often as 110 a second. Another group of flies which are more closely related to one another than they are to *D. melanogaster* and its relatives, is the obscura species group. These tend to produce longer notes of up to seven sine waves with a higher frequency of about 500 Hz. The repetition rates of these sounds varies from five per second in *D.*

pseudoobscura to twenty per second in its sibling species *D. persimilis* (Figure 2).

In *D. melanogaster* and *D. simulans* and a few other species, one wing is held out at right angles to the body during vibration while in most species including *D. persimilis* the wings are extended to a lesser extent. It is likely that by progressively unfolding the wing the fly can alter the extent to which the wing is coupled to the thoracic box thus changing the resonant frequency of the thorax and consequently of the sound produced.

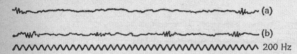

Figure 2 Courtship sounds of (a) *D. pseudoobscura* and (b) *D. persimilis* showing the same notes produced at different intervals

The fly thorax and wing form a mechanical unit when the wing is fully extended. The thorax is an elastic box with the peculiar property of holding the wing in two stable positions, up or down, but not between the two. If the raised wing is pushed downwards it will suddenly click into the down position and vice versa. In flight this system is used to maintain wing oscillation by producing suitable mechanical loading for the large flight muscles. The wing-beat frequency during *Drosophila* flight is about 200 Hz and recordings show that a fairly complex pattern of sound is produced. One of the components of this sound is virtually identical to the courtship note of *D. melanogaster* and *D. simulans*. We believe that the songs in these species are produced by repeated up and down movements of the wing, but with the main flight muscles unstimulated so that the entire resonant system does not go into oscillation. In *D. pseudoobscura* and *D. persimilis*, which produce songs with their wings partially folded, it appears that the main flight muscles are activated intermittently. The sounds produced, however, are of a higher frequency than in flight as the mechanical loading of the thorax by the wing is less when the wing is in the unextended position.

Unfortunately, the problem of insect flight is a very difficult one, especially with minute insects, and our study of this aspect of the problem has only just begun.

Various insects have organs that may be described as ears, with tympanic membranes. Grasshoppers have ears on the fore legs and, incidentally, have elaborate love songs and similar problems of finding the right mate as does *Drosophila*. Noctuid moths have thoracic ears used to detect the radar-like emissions of bats that prey on them. Flies, however, do not have ears with tympanic membranes.

Very recently a colleague, Aubrey Manning, has shown that if the feathery part of the fruit fly's antenna, the arista, is removed or glued down, female *Drosophila* became unreceptive. We have watched the antennae of *Drosophila* when loud pure sounds are played and found that the arista and distal joint of the antenna vibrate with a natural resonance of about 200 Hz. This suggests that the antenna is used to receive the love song.

Drosophila courtship is interesting at various levels. The love songs are produced by the wings in what appears to be a simple modification of the flight mechanism and the receptor system in females is adapted from one used in the control of flight. The sound language may have evolved from the buzzing noise that flies and other insects make when warming up before flight. A few species appear to have lost the ability to produce sounds and the wing displays provide visual stimuli.

Any language is restricted by the number of words in its vocabulary; this one is interesting in having only a very few words, but so far the only species with identical songs are two morphologically rather dissimilar species, *D. pseudoobscura* and *D. ambigua*. The former species is restricted to the North American continent and the latter to Europe and so are never likely to meet and hence need a language barrier. In contrast *D. pseudoobscura* and its sibling species *D. persimilis*, whose habitats overlap, have quite distinct songs. These two species are so similar morphologically that the most certain means of distinguishing between them is by listening to them courting.

The female fly also has a language but it consists of only one word: a long, very loud and very effective buzz. Its meaning to males of all species is 'No'.

H. C. Bennet-Clark and A. W. Ewing 281

23 M. P. Harris

Species Separation in Gulls

M. P. Harris, 'Species separation in gulls', *Animals*, vol. 12, 1969, pp. 8–12.

Gulls are among the most widespread, most numerous, and most obvious groups of birds. In Britain two of the commonest species – the herring gull (*Larus argentatus*) and the lesser black-backed gull (*Larus fuscus*) are of special interest to zoologists as, though very closely related, they rarely interbreed. Despite their different plumages and colours of the soft parts (a lesser black-backed gull has dark grey mantle and wings, yellow legs, red eye-ring and edge to the gape of the bill, as compared to a herring gull's pale mantle, yellow eye-ring, and flesh-coloured legs), some scientists have considered them to be the same species as they are at the ends of a chain of intergrading species of gulls circling the northern hemisphere.

In such difficult cases as this, any decision to call a population a species or a subspecies is purely arbitrary, but I prefer to call these two species – if only because they look different and rarely interbreed. In Holland, however, a very small number of cross-matings, giving rise to fertile offspring, have been observed over many years. These have been in colonies of herring gulls with a few lesser black-backed gulls, so possibly such matings occur only when a bird is unable to obtain a partner of its own species.

During six seasons spent studying many thousands of pairs of lesser black-backed and herring gulls on Skokholm and Skomer Islands (off the Welsh coast), I have never seen a normal herring gull paired with a lesser black-backed or hybrid gull, so obviously there must be some efficient mechanism for preventing interbreeding.

The behaviour of the two species is very similar, and differences are unlikely to be enough to bring about this separation. The two species have different calls, but there is so much variation between

individuals in either species that this again is unlikely to be the critical factor.

As far as food and habits are concerned the species are distinct. At present the herring gull obtains a large proportion of its food either directly or indirectly from man – in the form of fish-waste, offal and garbage; whereas the lesser black-back is a more natural feeder on shores and in fields. Although the latter species does feed on waste tips, especially outside the breeding season, it does not frequent fish markets.

Herring gulls on Skokholm and Skomer nest mainly on the cliffs and have the peak egg-laying time at the beginning of May. The lesser black-backs prefer the bracken-covered areas on the flat tops of the islands, and lay about ten days later than the herring gulls. There is, however, considerable overlap, and in some other colonies the two species nest side by side in a rather uniform habitat. Ecologically the differences are probably insufficient to prevent interbreeding.

The British herring gull is a sedentary species, and it is only rarely that birds move more than 240 km from where they were hatched. In contrast, most lesser black-backed gulls spend the winters in Spain, Portugal or on the north-west coasts of Africa. Herring gulls ringed as young in Britain have been recovered in these areas, but these birds have always originated from colonies where lesser black-backs are also present. As the young of the two species are very similar, most of these recoveries doubtless relate to lesser black-backed gulls misidentified at the time of ringing. On very rare occasions, however, young gulls are known to have been adopted by strange adults, and I wondered how any herring gulls reared by lesser black-backed gulls might behave.

In 1962 (on Skomer) and 1963 (on Skokholm) I interchanged all the eggs in several lesser black-backed-gull colonies with eggs in herring-gull colonies. The eggs were hatched and the young reared successfully. In all I ringed 407 herring gulls which had been reared by lesser black-backed gulls, and 335 young lesser black-backed gulls reared by herring gulls. Presumably any gull is imprinted on the adults it sees when it hatches, and so young gulls should 'think' that they look like their parents. However, birds reared by foster-parents of a different species would mistakenly 'think' that they looked like their foster-parents. (It

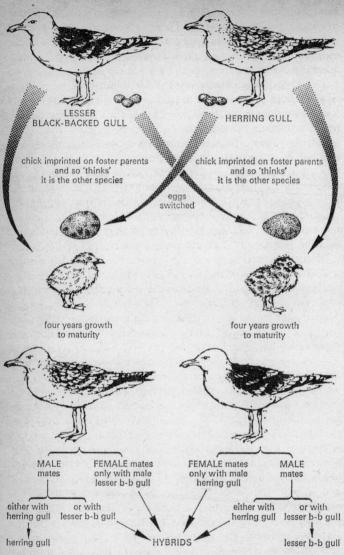

LESSER BLACK-BACKED GULL

HERRING GULL

chick imprinted on foster parents and so 'thinks' it is the other species

chick imprinted on foster parents and so 'thinks' it is the other species

eggs switched

four years growth to maturity

four years growth to maturity

MALE mates

FEMALE mates only with male lesser b-b gull

FEMALE mates only with male herring gull

MALE mates

either with herring gull

or with lesser b-b gull

either with herring gull

or with lesser b-b gull

herring gull

HYBRIDS

lesser b-b gull

Figure 1

should perhaps be stressed that although this is probably true for many birds, it could not hold for parasitic species whose young never see the parents. Otherwise a cuckoo raised by, say, a hedge sparrow, would try to mate with a hedge sparrow and not with another cuckoo.)

These egg-swapping experiments soon yielded results, as some of the ringed gulls were recovered away from Skokholm. As I had hoped, herring gulls were reported from the Bay of Biscay, Spain and Portugal: places where authentic British herring gulls are never recovered. The British population of herring gulls is unusual in being sedentary, whereas many other populations migrate. Perhaps the urge, and the mechanism for migration, still lies dormant in British birds and can be triggered off by the example of the foster parents, or by something in their up-bringing.

It was even more surprising that some of the lesser black-backed gulls were also recovered in southern Europe, as these – unlike the herring gulls – could not have followed their foster parents which would have stayed near the breeding colonies.

Further results of the switches were not obtained until 1968 when I was able to return to Skokholm to look for the resulting gulls. To my amazement I discovered no less than twenty-nine pairs of gulls where one partner was a lesser black-backed and the other a herring gull. All were in small colonies where I had interchanged the eggs. A short visit to Skomer showed two mixed pairs. Some of these pairs successfully reared hybrid young which were marked as a basis for future work.

Of the birds in these pairs, twenty-five were caught and shown to have been ringed as foster young; some others were seen to be ringed but could not be caught; some others were unringed and could have been normally reared birds (though none of the 1000 normal gulls reared every year were retrapped), or fostered birds which had not been ringed due to doubt as to their true parentage. Alternatively, they could have lost their rings.

In four pairs where both adults had been involved in fostering, the male was a lesser black-backed gull and the female a herring gull. This was surprising as the lesser black-backed is the smaller of the two species, and in a pair of gulls the male must be larger and have a heavier bill than the female; hence in these pairs

either the lesser black-backed male was very large, or the female was a very small herring gull.

As field work progressed and more mixed pairs were discovered, I thought that the explanation was straightforward: simply that the birds were choosing mates similar in appearance to their foster parents. However, these thoughts were shattered when I started to find fostered birds with mates of their *own* species. I dismissed the first few as errors in my original work but by the end of the season there were twenty-five such birds – so some other reason had to be looked for. It seemed unlikely that these young had not been imprinted on their foster parents while other birds had. Similarly, it was unlikely that so many birds would have made a wrong choice of mate, as if this was so many mixed pairs would occur naturally.

Examination of the sexes of these fostered birds led to an interesting explanation. It appeared that male birds will accept as a mate a female of either species; but a female will only choose a mate of her own species. Fostered females would then mate only with males of the other species. This theory was supported by an example of a female lesser black-backed, mated with a herring gull, which deserted her eggs when trapped.

Allowing that at least female gulls can recognize birds of their own species, what are the important recognition marks? Herring and lesser black-backed gulls are quite distinct, but this species separation is found in other areas where very similar-looking gulls nest side by side. Neal Smith, in an extremely interesting monograph published by the American Ornithologists' Union, has demonstrated that the eye-ring colour is very important. By altering the colour he was able to prevent birds of similar species from forming pairs, and even to induce some mixed pairs. Eye-ring colour, which is strikingly different in the herring gull and lesser black-backed gull, probably serves a similar function. As yet I have only been able to change the eye-ring colour in pairs which had already laid, and this had no effect – even if the original nest was destroyed. Probably by this stage the birds were recognizing each other as individuals and the species-separating marks were no longer important.

I had some misgiving about designing experiments to interfere with nature, in case any irreversible changes were brought about.

The present experiments, if carried out on a large enough scale, could theoretically bring about the fusion of two species into one, and serious thought was given to the desirability of allowing any hybrid young to be reared. As hybrids rarely occur in Europe – the two species remaining separate – there would seem to be some definite disadvantage in hybridization, making it unlikely that populations of hybrids would become established. But the experiments have been allowed to continue for another year, as the importance of results to be gained outweighs any slight risk involved.

24 D. Jukes

Testing Time for Tongue-Twisters

D. Jukes, 'Testing time for tongue-twisters', *Science in Action*, vol. 1, 1969, pp. 584–5.

The ability of some people to roll the tongue into a U-shape has been known to geneticists for many years. In a survey conducted by geneticists in Baltimore, USA, 5000 people were examined for the trait. After all the results were in, it was found that 3200 could roll their tongue, whereas 1800 people could not. This represents a total of 64 per cent rollers.

Further surveys showed remarkable resemblance – i.e. 64 per cent rollers. In fact, the percentages obtained were within 0·1 per cent of the Baltimore results. This was too accurate to be purely coincidental, and geneticists classed the phenomenon as simply a *Mendelian dominant*. On human chromosomes there are pairs of genes – the units responsible for transmitting to future generations hereditary traits such as hair texture, height and eye colour.

Let us suppose that there is a pair of genes that transmit the tongue-rolling ability. If r stands for the recessive (i.e. non-rolling), and R stands for the dominant (rolling) ability, a person can have three different gene combinations or *genotypes* – RR, Rr, rr. The first two are both dominant, the first being homozygous dominant, the second heterozygous dominant or hybrid, the third recessive.

The whole question of tongue-rolling at this stage seemed to me to involve more than the geneticists might have us believe. Surely the muscles in the mouth must have some effect. Secondly, it amazed me that people, including myself, who at first count were classed as non-rollers, could train themselves and soon become proficient rollers. This, however, still did not upset the genetic equilibrium.

The County High School at Arnold in Nottingham conducted a survey, and the results they obtained for percentage rollers and

non-rollers were once more identical to the original results. They theorized that people breast-fed in infancy could roll their tongues.

I decided to carry out a large-scale survey in Wolverhampton, and with the help of my biology master, Mr C. Westall, I had over six hundred three-page questionnaires duplicated.

Each questionnaire had space for the data of six people – i.e. one questionnaire per household. The questions I asked were as follows. (1) Can you roll your tongue? (2) Sex (to determine whether there was any link between this and the tongue-rolling ability). (3) Method of feeding in infancy, i.e. breast-fed, bottle-fed or a combination of the two. (4) Did you suck your thumb, finger or a dummy in infancy?

During August 1968, I circularized the questionnaires and obtained the cooperation of over 1000 people in the area. The response was naturally mixed – for, after all, the question 'Can you roll your tongue?' seems rather bizarre to the uninitiated.

I also traced back the tongue-rolling family tree of each family for three generations. This, I hoped, would finally prove or disprove the argument for and against the trait being hereditary. If a marriage between a pair of non-rollers (which must have the genotype rr) produced children with a tongue-rolling genotype then, allowing for a possible small proportion of throwbacks, this would tend to suggest that genetics has nothing to do with tongue-rolling.

The results that I obtained for the roller/non-roller percentages were once more identical to the previous results: homozygous dominant 15·8 per cent, hybrid dominant 47·9 per cent and recessive 36·29 per cent.

The data from the second stage of the survey I transferred to computer cards, and with the help of Mr John Kwok, a computer programmer at the Wolverhampton College of Technology, I was able to write a program in Fortran (a particular computer language) for the survey. By feeding this to the IBM 1620 computer at the College of Technology I was able to correlate the following information. Twice as many of the people that were breast-fed in infancy had the tongue-rolling ability. I also found that in twenty-seven crosses between non-rollers, sixteen did not breed true – they produced children who could not roll their tongues.

The conclusions that I draw from this are as follows. I believe that environment plays a part in tongue-rolling, for twice as many rollers were breast-fed. There are in my opinion two possible explanations for this. Firstly, at birth rollers possess the genotype for rolling and those who are breast-fed are more likely to develop the rolling *phenotype* (a phenotype is a type produced by the reaction between a given genotype and given living conditions). The muscle actions in breast-feeding might stimulate the development of the rolling muscles. Non-rollers are fed at first at the breast, but their lack of these muscles leads to difficulties and they are switched to the bottle to alleviate them.

I do not disagree with the hereditary aspect of tongue-rolling. I am merely trying to assess critically whether it is as simple as a Mendelian dominant. Incomplete dominance is a general principle which I think might well apply. When a black person marries a white person, then the resulting offspring are neither white nor black – they are in fact of an intermediate shade. When a blue-eyed person, however, marries a pure homozygous brown-eyed person, then the offspring are always brown, allowing for throwbacks etc. In both of the above cases the darker colour is said to be dominant over the lighter colour. In the first case, however, the dominance is referred to as being incomplete. In simple terms, this means that the genes for the lighter colour are not completely 'masked' by the genes for the darker colour.

This could well be the case in tongue-rolling. In other words, there is an imaginary scale ranging from roller to non-roller. This would explain why the non-rollers are not a uniform group – they do not breed true.

Many of the so-called non-rollers may well in theory possess a small amount of the rolling genotype, but not enough to show the trait in the phenotype. However, when two such non-rollers cross, then the strength of rolling genes present is increased and it is then feasible for offspring to have the rolling phenotype.

The most intriguing question to arise from the survey was: 'Has tongue-rolling any survival value for today?' This, however, was not always asked from the evolutionary and biological point of view, but from the angle of waste of time and money researching into an obscure subject such as this.

Many people did consider it from the biological aspect. Was

this, for example, a form of discontinuous variation? Was the frequency of rollers and non-rollers constant in different races? A possible hypothesis is that the ability to pronounce certain phonemes, which is controlled largely by the tongue, may well be affected by the rolling ability. In speech therapy the tongue is rolled around a pencil. It may follow that various languages have evolved from the ability of the indigenous populations to roll their tongues. But this is purely hypothetical.

In retrospect, I feel that it would have been valuable to have composed an additional question on the questionnaire referring to race, comparing Australoid, Caucasoid, Mongoloid and Negroid. This is one reason why I feel that it would be of great value for other schools, not only in this country, but also abroad, to carry out similar surveys. The research into such a simple and obscure phenomenon as tongue-rolling is endless.

25 B. W. Thompson

Let 'em Roll!

B. W. Thompson, 'Let 'em Roll!' *Science in Action*, vol. 1, 1969, p. 688.

The questionnaire which Douglas Jukes used in his Wolverhampton survey [Reading 24] on the ability of people to roll their tongues into a U-shape was devised by the pupils of the Arnold County High School, Nottingham, in an attempt to explain why some non-rollers with practice could become rollers. We reasoned that if they had the necessary genotype but failed to develop the trait it must be due to a lack of stimulation, and since the tongue is involved in feeding, we looked for a difference in feeding common to all humans. This applied only during baby-feeding, and we therefore decided to survey a sample of the Arnold population according to their tongue-rolling abilities and how they were fed when babies. We also questioned them about blood groups, sex, right- and left-handedness, thumb-sucking and use of comforters, to see if other factors might be involved.

Our sample of 720 produced the following data and conclusions.

Of the sample, 90 per cent were right-handed, 55 per cent were breast-fed, and 46 per cent sucked either their thumb or a comforter.

There is no link between right-handedness and tongue-rolling; sex and tongue-rolling; sucking thumbs and comforters and tongue-rolling; or blood groups and tongue-rolling.

There *is* a definite connection between tongue-rolling and breast-feeding.

In 1951 Taku Komai surveyed the Japanese and found that there is a link between tongue-rolling and sex. The incidence of non-rollers is always greater in men than women (29 per cent of men are non-rollers and 25·5 per cent of women). Also the percentage of non-rollers changes from 50 per cent in seven-year-old

boys to 29 per cent at twelve, after which it remains fixed. This trend also occurs in girls, but stabilizes at a lower percentage.

J. Warren Lee in 1955 sampled 1890 Negro college students and discovered that 82 per cent were rollers. Thus it would appear that genetic and environmental factors are responsible for variations within racial groups. European communities have 64 per cent rollers.

We now began to wonder whether the link between tongue-rolling and breast-feeding was causal or coincidental. It was obvious that the hypothesis required a different muscular action of the tongue in breast-feeding as opposed to bottle-feeding. We therefore asked Dr Mavis Gunther, of University College Hospital, London, whether a non-roller would be unable to breast-feed successfully and would therefore be switched to the bottle; or whether both could suckle successfully, but a tongue-roller would be able to exercise the muscles during feeding.

It emerged that bottle-feeding does demand a different technique. The mouth is held in a small circle, and the tongue is kept well up, to guard against having the teat pushed in too far. This difference is so noticeable that Gunther can tell a bottle-fed baby in this way within 2–3 days of its being switched from breast-feeding to bottle-feeding. During breast-feeding the nipple is held in the baby's mouth by suction produced by the back of the tongue, but Gunther thinks it probable that the front part of the tongue may curl round to keep a firm seal. The main body of the tongue functions similarly in both types of feeding; it has a progressive ripple in its action, rather like the pulsatile flow pumps developed for the heart–lung machine. In this way the milk is expressed from the bottle teat or from the ducts behind the nipple and the areola.

We can therefore discount the theory that tongue-rollers would have been at high survival value in days before feeding bottles. But breast-feeding might encourage the development of the rolling muscles if the genotype was favourable.

The next key question was: is man the only species of tongue-rolling ape? Can apes roll their tongues? We wrote to Dr Fae Hall, the Information Officer of London Zoo, to ask her if she could investigate the apes on our behalf. Outside infancy she concluded that apes were non-rollers, since they have the *frenulum*

linguae attached near the tip of the tongue, so that they have only a small part of the tongue which can be protruded. When she offered juice on her finger, the chimps, orangs and adult gorillas all held their tongues flat. The suckling of infant apes is impossible to observe close to; but it seems unlikely that they do use the curling action around the teat, because they have immensely powerful and protuberant lips, and these may form the seal. The lower lip is curled into much the same shape as a tongue-rolled tongue.

But we cannot say that man is a unique tongue-roller from this data, because we cannot ask apes to roll their tongues. This brings us to the crux of this problem: is tongue-rolling a natural tongue movement at all? It may be in infancy, but outside that it never occurs as a spontaneous movement either in speech or play. Dr Burton Jones, who has studied the behaviour of nursery children for some years, says that although they display great versatility on tongue movements, he has never observed spontaneous rolling.

If we consider the mechanics of tongue-rolling, it is inconceivable that a pair of *allelic* (different) genes could determine the ability to roll or not. The muscles of the upper lip must play a key part. The muscles in the tongue bring about first a broadening and flattening, then protrusion, and finally the operation of many muscles to form a hollow tube. The inheritance must be *polygenic* (derived from many genes), and may involve muscle groups developing, or nerve pathways being established with the key muscle groups by the lingual nerve.

This is probably the reason why non-rollers do not breed true. Each one lacks a particular gene, which provides a key part of the mechanism that is different in both parents. The offspring may have the full complement of genes and are therefore rollers.

If the trait is polygenic in inheritance, it is likely that the genes concerned are linked with other genes. A useful survey would be to determine whether there is any correlation between tongue-rolling ability and body build (using an equation which combines stature and weight).

Finally, it would be valuable evidence for or against the hypothesis of a causal link between breast-feeding and tongue-rolling to carry out a survey in which parents and offspring were assessed

separately. Bottle-feeding has increased enormously over the past twenty-five years, and if there are fewer rollers among children between twelve and twenty compared with people between thirty and fifty, this must indicate that an environmental factor is at work, and breast-feeding or bottle-feeding is the likely answer.

26 R. Riley

Plants: Off-the-Peg or Made-to-Measure

R. Riley, 'Plants: off-the-peg or made-to-measure', *Advancement of Science*, vol. 24, 1967, pp. 217–24.

Initial pattern of plant improvement

Something like ten thousand years ago our ancestors first began to adopt a settled agricultural life instead of the wandering existence they had previously followed as hunters and gatherers. Of course the change to a predominantly agricultural life must have been gradual – occurring in some areas earlier and in others later. In addition, total dependence upon the harvest of cultivated crops, rather than upon the gathered produce of untended plants of forest and plain, was probably only accomplished over an extended period. Indeed even now in Britain we have not totally abandoned the gathering of wild produce. The delicious bilberry pies cooked in local Yorkshire kitchens, from fruit picked on the Pennine moorlands, testify to the continuing usefulness of plant produce collected from the wild. Elsewhere present-day gatherers seek out blackberries, sloes, rose hips, elderberries and a host of other fruits, while the blueberry provides a national dish for our American friends. However, although we have not entirely abandoned gathering wild produce, there are now few people anywhere in the world who obtain their staple foods from the wild.

The production of food crops continues to be, as always, our most important industrial activity. The nature of this activity is determined by the kinds of crops bequeathed to us by our ancestors and by the ways in which we can modify these crops better to satisfy our needs. Some plant species more or less surrendered themselves into agricultural bondage by the freedom with which they grew in the disturbed areas round human dwellings. However, relatively few species were chosen to become crops from the profusion of nature, by the early agriculturalists. The number was

limited because few species that were amenable to cultivation provided useful produce that was easily harvested and stored.

These were the plants that, in the metaphor of my title, were taken off-the-peg of nature. Immediately that they were taken into cultivation man began by unconscious selection to tailor them to fit his needs better. The mere acts of sowing and reaping are selective and, as a result of these and all the basic farming practices, genotypes better suited to the conditions of agriculture were favoured in the initially heterogeneous populations. Crop plants, more or less as we know them today, emerged from several millenniums of such unintentional selection. In the process they often became so different from their wild progenitors that they were incapable of persisting out of an agricultural environment.

Selection took place not only within but also between species and some, such as *Chenopodium album* and *Polygonum lapathifolium*, were discarded and are no longer cultivated. Possibly such species were rejected because they responded inadequately to agricultural selection. Certainly plants that survived for long as crops were shaped to meet man's needs by the selective favouring of genotypes that gave more – or more certain – yields, and that were better suited to agriculture.

While the initial shaping of the genetic fabric of crops species resulted from unintentional selection – and this process is still significant in moulding crops – conscious acts of choice became significant probably quite early in the development of agriculture. Obviously we can only guess at the rigour of the selection practised by our early farming ancestors, but it seems likely that it was severe because crop plants evolved rapidly. In parts of the world where primitive forms of agriculture are still carried out, extreme care is apparently taken over the source and choice of seeds for the next crop. Stringent selection is possible because of the close observation of, and feel for, crops. Detailed, almost individual, knowledge of his plants is feasible for a farmer working under primitive conditions in a way that the modern grower, tending large areas by machine, would find it hard to imagine. Much of the tailoring of crop plants to their present forms was accomplished in this apparently crude manner, and the more we consider the extent of the adaptation of our crops to cultivation the more must we wonder at its effectiveness.

R. Riley 297

Much more deliberate selection was carried out in the eighteenth and nineteenth centuries, especially following the spread of understanding of the theory of natural selection. From these activities selected varieties arose that were genetically purified, multiplied and widely disseminated. A good example of this process [. . .] was narrated by the famous nineteenth-century selectionist S. D. Shirreff. It concerns the establishment of the wheat variety Squarehead which originated from a single prolific and high-yielding plant spotted in a field of Victoria wheat in Yorkshire in 1868 by a man called Taylor. From 1870, Squarehead was grown and sold by C. Scholey of Goole, who may in part have been responsible for its discovery. By the 1880s Squarehead had spread to Scandinavia, France, Germany, Holland and Belgium. This illustrates the way in which ease of transport enabled talented and creative selectionists – able to pick out promising variants – to exercise influence over crops well beyond their own localities. Much of the detailed shaping of the crop forms used today resulted from the efforts of these selectionists.

Finally, in outlining the history of the development of plant improvement, mention must be made of hybridization. Breeding by hybridization and selection has been carried out in many of our most important crop species for a little over a hundred years. In wheat, for example, the first attempts at hybridization were made by Knight at the end of the eighteenth century and wheat hybrids were demonstrated at the Great Exhibition of London in 1851.

Hybridizers of cultivated plants make use of rearrangements, in the derivatives of their programmes, of the characters that distinguish the parental forms. They also look for expressions of beneficial characters that surpass those of either parent. In the segregating generations following hybridization they select, for further propagation, plants with arrays of characters that approximate to the ideal they have set themselves. This fashioning of a more acceptable pattern of plant depends upon recombination and segregation of the chromosomes of the initial parents at meiosis in the hybrids and their derivatives.

Great progress in the genetic improvement of cultivated plants has resulted from the complementary procedures of hybridization and selection and elaborate methods have been designed to

increase the probability of obtaining the desired variants from such programmes. Much current plant breeding makes use of these painstaking procedures and a good deal of the steady rise in arable productivity derives from their application. However, since they are fairly well understood I will not discuss them further, concentrating instead upon some more unusual and complex approaches.

In its pursuit of more valuable cultivated forms, plant breeding is not concerned with the enunciation of new principles from nature, but – in a broad sense – with the modification of our environment; so it is not science but technology. However, it employs knowledge gained from numerous scientific disciplines, and many plant breeders are forced to apply scientific method in order to provide themselves with the knowledge necessary for the attainment of specific breeding objectives. The ever-increasing use of scientific understanding, and the employment of refined techniques derived from science and mathematics, are enabling plants to be modified in ways previously impossible. In order to help us to understand this – and to illustrate our improved capacities to make plants to fit our measurements – I will outline a few examples.

Fashioning the product

A great deal of current work on crop improvement has been stimulated by the more precise definition of the use to be made of the harvested product. Some definitions have been formulated because of extensions of the manner in which crops are used. For example, we no longer simply require peas, – but peas suitable for this freezing or that canning process; no longer potatoes but potatoes for crisping, for chipping and freezing, and so on. This diversification in use is being met by the development of genotypes yielding products that fit the various user requirements.

Of much the most profound significance, in terms of the adjustment of crop products to user requirements, are the recent discoveries concerning the inheritance of amino acid composition in maize. In a major scientific breakthrough Mertz, Bates and Nelson, of Purdue University, showed that homozygosity for either of the mutant alleles opaque-2 (o_2) or floury-2 has the effect of approximately doubling the lysine and tryptophan content of.

maize grains (Table 1). The significance of this discovery is that the low concentrations of these amino acids normally limit the nutritional value of maize to monogastric animals such as men and pigs. The nutritional advantage of *opaque-2* over normal

Table 1 Amino Acid Content of Normal and *opaque-2* as g/100 g of Protein

Amino acid	Normal	Opaque-2
Lysine	1·6	3·7
Tryptophan	0·3	0·7
Histidine	2·9	3·2
Arginine	3·4	5·2
Aspartic acid	7·0	10·8
Glutamic acid	26·0	19·8
Threonine	3·5	3·7
Serine	5·6	4·8
Proline	8·6	8·6
Glycine	3·0	4·7
Alanine	10·1	7·2
Valine	5·4	5·3
Cystine	1·8	0·9
Methionine	2·0	1·8
Isoleucine	4·5	3·9
Leucine	18·8	11·6
Tyrosine	5·3	3·9
Phenylalanine	6·5	4·9
% Protein	12·7	11·1

Maize grain is illustrated from an experiment carried out with rats by Mertz and co-workers in Figure 1. Similar benefits occurred when *opaque-2* maize was used to feed children.

I am sure that in this hungry world I need not labour the advantages of removing the nutritional limitations to the usefulness of maize that are consequent upon the use of such mutants. High-lysine maize stocks, with high yields, high overall protein contents and appropriate combinations of agronomic characters, are now being bred in the USA. But it is in some of the underdeveloped countries – where malnutrition is common – and where maize is used directly as human food that the discoveries made at Purdue

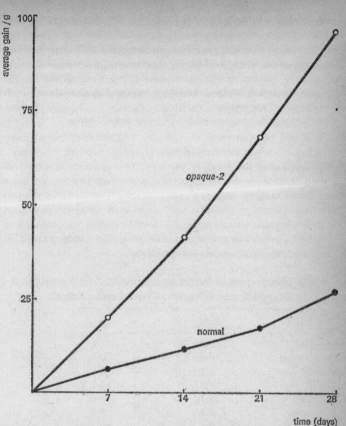

Figure 1 A graph to show the average weekly gains of weight of rats fed on *opaque-2* high-lysine maize or normal maize

University will have the greatest impact. In this case, making the crop measure up to our needs may be one of the most useful contributions that technology can make to the food and population problem.

As a footnote to these comments on high-lysine maize it is sobering to recall that the *opaque-2* and *floury-2* mutants have been known to maize genetics for about forty years. The recognition of their true significance and potential value required the

perception of Dr Nelson and his colleagues and the availability of amino acid auto-analysers. Searches are now being made for similar genetic variants in other cereal species including the temperate cereals, wheat and barley. Success in these searches could also have profound effects in correcting disequilibria in the world's nutrition. Indeed, breeding plants with nutritionally advantageous proportions of amino acids may be the most important detailed fashioning that our crops can undergo.

It might be interesting to consider another example of a detailed adjustment of crop products made possible by greater appreciation of the needs of the user. This concerns the oil produced from the seeds of rape (*Brassica napus* $2n = 4x = 38$), a rapidly expanding crop in Canada. Rape oil is used as an industrial oil, when high linolenic and erucic acid contents are valued. It also has a use as an edible oil – for margarine, shortening and salad oil – when low erucic and eicosenoic acid contents and high oleic and linoleic contents are preferred.

Table 2 Comparison of Percentage Fatty Acid Composition in Seeds of Rape (*B. napus*) Varieties Tanka and Canbra

Fatty acid	Tanka	Canbra
Palmitic	3·7	4·8
Palmitoleic	0·4	0·5
Stearic	1·2	2·4
Oleic	15·7	63·1
Linoleic	15·2	19·4
Linolenic	9·3	9·0
Arachidic	0·9	trace
Eicosenoic	9·2	0·8
Erucic	44·4	0

Little progress could be made in the adjustment of the fatty acid content of rape seed by breeding until rapid chemical methods became available that permitted large-scale screening. However, with the application of gas chromatography Downey and his colleagues at the Canada Department of Agriculture Research Station, at Saskatoon, found that they could easily select for fatty acid content. Determinations were made from one

cotyledon removed from the embryo and the remainder of the seed was germinated and planted. In this way the zero erucic acid variety, Canbra, was quickly established. Moreover, in Canbra, other favourable changes in fatty acid composition were the low level of eicosenoic acid and the high levels of oleic and linoleic acid contents, although selection had not been exercised directly for these components (Table 2).

Canbra, which was tailored to their needs, is now benefiting margarine and salad-oil processors. In addition, studies of the inheritance of erucic acid content have shown that it is probably controlled by two independent loci (Table 3). Consequently, it

Table 3 Genotypic Constitutions of *B. napus* Postulated to Determine Various Levels of Erucic Acid Content

% Erucic acid	Genotype	
0	aa	bb
	Aa	bb
9–10	aa	Bb
	AA	bb
18–20	Aa	Bb
	aa	BB
27–30	Aa	BB
	AA	Bb
36–40	AA	BB

seems that erucic acid content could be stabilized at the 0 per cent, 18–20 per cent or 36–40 per cent levels – the upper range providing oils suitable to industrial use. This example illustrates the precise adjustment of a product to the requirements of its processors that becomes possible when applied genetics is sustained by rapid methods of chemical assessment.

'Instant' crops

So far, I have emphasized the gradual nature of the evolution of crop plants and indicated the detailed refinements that can result from the activities of plant breeders. Now I would like to discuss an example where the scale both in terms of the speed and of the magnitude of change is a complete contrast. This concerns

combinations of wheat with rye that have recently been interesting cerealists.

The most widely cultivated wheat species is *Triticum aestivum* – the bread wheat – an allohexaploid with forty-two chromosomes. Allotetraploid wheats with twenty-eight chromosomes, such as *T. durum* – the macaroni wheat, are also of considerable importance however. Cultivated rye, *Secale cereale*, is a diploid with fourteen chromosomes and for long cereal breeders have cherished the notion of combining the high productivity, and grain quality, of wheat with the vigour and hardiness of rye. This resulted in the extensive study of 56-chromosome synthetic allo-octoploids derived from the cross, *T. aestivum* × *S. cereale*. However, in these forms the genetic complements of the parental species do not combine to operate in an integrated manner, so that despite extensive breeding work no worth-while crop has been developed.

In the last decade interest turned instead to 42-chromosome allohexaploids derived from crosses like *T. durum* × *S. cereale*. The chromosome number of this so-called *Triticale* species is not new since it is the same as that of bread wheat. Hybridization between *Triticale* forms of independent origin, followed by selection, has resulted in the production of genotypes that are at least of sufficient merit to demand further work.

A variety of *Triticale* bred by Kiss is now grown on a farm scale on the very sandy soils of the Kecskemét area of the Hungarian Plain. It replaces rye as a higher-yielding cereal species, tolerant of low-fertility conditions. To Kiss must therefore be given the credit of being the first person in historical times to bring a new cereal species into cultivation. In addition, however, workers at the University of Manitoba, Winnipeg, are using the hexaploid *Triticale* in breeding, and their varieties are also grown on a farm scale in southern Manitoba.

A unique interest of some workers with *Triticale* is the provision of a new type of grain from which whisky can be distilled. This is of obvious concern in regions where rye whisky is favoured, and many thousand of proof gallons of *Triticale* whisky are maturing at the present time. Proof of the whisky will be in the tasting, and seven years is a long wait, but the future of *Triticale* does not depend upon this alone. Its future will ultimately be determined by its yields relative to those of the other temperate

cereals, by its hardiness, vigour and disease resistance, as well as by the quality of its grain, of which whisky flavour is only one component. However, from what I have said about *Triticale* I hope that I have demonstrated that plant breeders are not only interested in making small-scale – although very important – adjustments to existing crops, but also concern themselves in the investigation of 'instant' crops.

The up and down of chromosome numbers

In order to adjust the structure of crop plants, or to make improvement easier, the plant breeder occasionally changes their chromosome numbers. The procedures involved employ principles derived from cytogenetics and indeed sometimes partly copy situations found in nature in other species. An example of this relates to the normal condition of bananas and an artificially maintained condition of some water-melons.

The fruit of the edible banana develops parthenocarpically, failing to set seeds for a number of reasons, one of which is that the most widely grown cultivars are triploid. The seedlessness of the fruit is of considerable importance in relation to the edibility of the pulp and triploidy plays a significant role in this character, although of course it also means that useful cultivars must be vegetatively propagated.

The water-melon (*Citrullus vulgaris* ($2n = 22$) is a native of tropical Africa although it is now widespread outside temperate and arctic areas. As you are probably aware, the numerous seeds do not detract from the enjoyment of eating water-melon but they can certainly make it inconvenient. It occurred to Dr H. Kihara, of the Japanese Institute of Genetics, that this inconvenience might be removed if triploid water-melons were produced. This notion, which relies upon employing genotypes analogous to those of banana, was so successful that triploid seedless water-melons are now produced in Japan, California, Taiwan (for export to Hong Kong) and Bulgaria (for export to central Europe).

The essentials of the system employed are that seed-producing, 44-chromosome, autotetraploid stocks are produced from normal diploids by colchicine treatment. Tetraploid–diploid crosses are tested to determine which parental combinations give good triploid offspring and seed stocks of the selected parents are

multiplied and grown together. Then tetraploid ♀ × diploid ♂ crosses are made on a large scale by hand-pollinations to produce seeds from which 33-chromosome seedless triploids are grown. By lifting the chromosome number to the triploid level, therefore, an alteration is made that considerably enhances the product.

Breeders of other crops have found a reduction rather than an increase in chromosome number to be valuable. This applies to certain work with the potato, *Solanum tuberosum*, which is a tetraploid species with forty-eight chromosomes. Because of its largely autotetraploid nature, seed setting is variable and often very low in the potato. Autotetraploidy and the accompanying low fertility makes new variation that arises by sexual reproduction difficult to exploit in breeding work. However, Hougas, Peloquin and Ross and their colleagues at the University of Wisconsin conceived the notion of circumventing this by reducing the potato to the 24-chromosome diploid (or polyhaploid) state. This was accomplished by making pollinations on to normal potatoes with pollen parents so marked genetically that seedlings could be recognized that derived from seeds that had developed without fertilization.

Haploids, produced in this way, in addition to displaying the simple disomic inheritance of diploids, are often vigorous and fertile. Consequently the potato can be bred as though it were a diploid and then restored to the tetraploid level by colchicine treatment if this is essential to obtain agriculturally adequate yields. However, even this is not certain, for it now appears that the potato is capable of producing acceptable yields of tubers at the 24-chromosome haploid level. Reductions in polyploidy are used only rarely in manipulating plant genotypes but the work with potatoes shows that it is an interesting means of shaping some crop species.

Imported characters

Occasionally wild relatives of cultivated species have genetic attributes that it would be useful to incorporate in the cultivated form. The plant breeder's task is then to transfer this variation and so to build up crops with characters imported from foreign sources. Such characters have been widely used and have contributed real benefit to the cultivated species.

I shall describe two examples of the transfer of useful characters between species to illustrate the kinds of procedures that are used. The first concerns the garden delphinium and is chosen in part for its aesthetic qualities although it is also of real merit as an exercise in plant breeding. The garden delphinium (*Delphinium elatum*) is a tetraploid species with thirty-two chromosomes, and its flowers range from white to deep blue in colour. Horticulturalists have been intrigued by the notion of obtaining red or yellow delphiniums especially as there are some related species with flowers in these colours. The related 16-chromosome diploid species *D. nudicaule* and *D. cardinale* are respectively orange- and red-flowered.

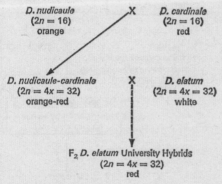

Figure 2 The origins of the red-flowered delphiniums produced by Dr R. A. H. Legro

The successful transfer of the red coloration of these species to garden delphinium was achieved by Dr R. A. H. Legro of Wageningen in the Netherlands. The procedure used is depicted in Figure 2 and involved first the hybridization of *D. nudicaule* and *D. cardinale*. The chromosome number of the hybrid was doubled using colchicine to produce a synthetic allotetraploid with orange flowers. The synthetic tetraploid was crossed with the natural tetraploid *D. elatum* and selection was practised for colouring, spike-shape and so on, in the generations derived from the resulting hybrids. After five generations of selection acceptable delphiniums with red and orange inflorescence were

obtained and are being described under the general name 'University Hybrids'.

The second example of alien variation transferred to a cultivated species, concerns the introduction by myself and my colleagues of disease resistance from the wild grass *Aegilops comosa* into bread wheat. One of the major diseases of wheat in Western

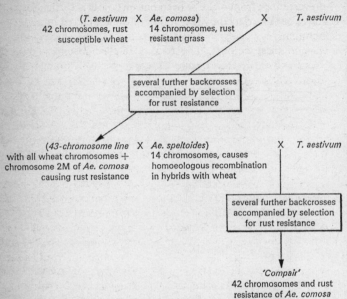

Figure 3 The origin of the wheat from 'Compair' which has the yellow-rust resistance of *Aegilops comosa*

Europe is yellow rust, caused by the fungus *Puccinia striiformis*. *Ae. comosa*, a 14-chromosome diploid from the eastern Mediterranean, is resistant to this disease and our task has been to transfer this resistance to wheat.

Hybrids, between wheat and *Ae. comosa*, and their derivatives were backcrossed with wheat, and rust-resistant derivatives, in each generation, were again backcrossed with wheat (Figure 3). The outcome of this programme was the isolation of a rust-resistant line that had the unchanged chromosome complement of wheat with, in addition, the single chromosome – designated

2M – of *Ae. comosa* determining rust resistance. From this work it could be concluded that in normal circumstances this *Ae. comosa* chromosome did not pair at meiosis and recombine with any wheat chromosome. Consequently, the region of the chromosome causing rust resistance could not be incorporated in wheat by conventional breeding methods.

From our earlier work with wheat we were aware that genetically related, or homoeologous, chromosomes are prevented from recombining because of the activity of a particular gene on chromosome 5B. We reasoned that, if the activity of this gene were removed or suppressed, the alien chromosome 2M would recombine with its homoeologues in the wheat complements and in this way the rust-resistance gene could be introduced into wheat. This plan was carried out using the genetic activity of another wild grass, *Ae. speltoides*, to suppress the activity of the 5B recombination-preventing gene. Consequently chromosome 2M, with its rust-resistance gene, was given the opportunity of recombining with wheat chromosomes.

Another backcrossing programme was carried out, again accompanied by selection for rust resistance, the outcome of which was the isolation of a line that, apart from its rust resistance, is perfectly normal wheat. Because it carries a pair of rust-resistance genes from *Ae. comosa* we call this line 'Compair'. In the construction of 'Compair', wheat parents were used that allow for a rapid turnover of generations but are not suited to farm use. However, the present situation is that 'Compair' can be easily used by breeders as a parent in the production of varieties that combine an alien form of rust resistance with other desirable agricultural properties. Thus a hole in the genetic cover of a crop has been patched with a gene from a wild species.

The plant tailor

I have discussed in outline a number of examples in which breeders have rejected the kinds of plants provided by nature or by their predecessors and have altered them to fit the needs of the grower or user. This is a technology in which success can be gained by the exploitation of understanding obtained from such sciences as agricultural botany, genetics, cytology and from plant evolution, chemistry and pathology. Important characteristics of plant

breeders are the optimism to believe that a worthwhile end will result from a prolonged programme of work and the persistence to carry through the programme. To those who successfully complete programmes of the kind I have discussed, however, comes the enormous satisfaction of having adjusted a small part of nature to their wishes and to the benefit or pleasure of the world at large.

27 S. R. Tannenbaum and R. I. Mateles

Single-Cell Protein

S. R. Tannenbaum and R. I. Mateles, 'Single-cell protein', *Science Journal*, vol. 4, 1968, no. 5, pp. 87–92.

In the remote Sahara, near Lake Chad, one of the staple human foods is a rich soup prepared from 'cakes' known locally as *dihé*. Each cake is a fibrous mass composed of millions of microscopic algae scooped as a green slime from brackish waters and dried in the sun. All over the world modern science is seeking to emulate the inhabitants of Chad and Niger and produce acceptable human food from micro-organisms rich in high-quality protein. The end product of this work is colloquially known as SCP, for single-cell protein.

SCP is a source of protein derived from such unicellular organisms as bacteria, yeasts, fungi, algae or protozoa. It can be grown on an appropriate culture medium, collected as cells and treated by various washing and drying steps to yield a powder consisting of dead cells containing 35–75 per cent protein suitable for use as food. The main advantage of SCP as a source of protein is that the organisms can be propagated on inexpensive substrates such as waste carbohydrate or hydrocarbons, requiring additionally only sources of nitrogen, phosphorus, magnesium, sulphur, trace elements, water and air. They grow much faster than our other food sources and convert raw material to final product more efficiently and so can produce a given quantity of food in a smaller space in a shorter time.

Nevertheless, only small amounts of SCP are at present being used and these mainly in the form of animal feeds and primarily as a source of vitamin B complex. The reasons for this are technological, economic, social and nutritional, as we shall explain.

There can be no single, simple solution to the world protein problem. SCP is but one potential source; others [. . .] include

oilseed-meal protein concentrates, high-lysine corn [see the previous reading], fish-protein concentrate, leaf-protein concentrate, fish farming and the domestication of new animal species. SCP must eventually compete with these in a more or less open market. Its advantages may be found to lie in flavour, acceptability, physical properties, cost or nutritional value.

A protein should contain the proper complement of amino acids required as external dietary factors by man. Of the twenty or so amino acids found in nature eight cannot be synthesized at the necessary rate in the human organisms and so must be available in the food supply. Of these, three are more often found to be limiting in human diets: lysine, methionine and tryptophan.

Unlike many plant proteins the lysine content of SCP is adequate and sometimes extremely high. This indicates the possibility of supplementing lysine-deficient proteins with SCP. The methionine and tryptophan content of most species is generally too low and variable; but the search for the ideal organism has only just begun, and SCP's inherent possibilities are evident from the proportions of lysine and methionine found in mixed rumen bacteria. That desirable lysine and methionine contents can be found in a single organism is shown by a strain of *Bacillus stearothermophilus* recently isolated in our laboratory at the Massachusetts Institute of Technology (see Table 1).

It would be obvious at this point that SCP could be used either as a complete source of protein or, more likely, as a protein supplement to other food sources. Thus, food staples such as wheat (deficient in lysine), soybeans (deficient in methionine) or corn (deficient in tryptophan) could be improved in protein quality. Micro-organisms can probably be found that either have amino acid contents close to the ideal, or two or three organisms which have complementary patterns can be mixed to obtain a product of high biological value.

Protein content is variable. It depends not only on species variations but also on cultural conditions. If bacteria are grown with limited nitrogen in the presence of an excess of carbon substrate they will tend to lay down energy reserves in the form of glycogen or poly-β-hydroxybutyric acid; yeasts and moulds, on the other hand, tend to lay down fats and other lipids. Either

Table 1 SCP sources

	protein per cent	lysine	methionine	tryptophan
		grammes per 100 g protein		
FAO recommended level		4·2	2·2	1·4
yeast { Candida tropicalis	45	7·7	0·8	0·8
Saccharomyces cerevistae	50	7·3	1·2	1·1
bacteria { Bacillus megaterium	40	7·0	1·8	0·6
Bacillus stearothermophilus	75	7·4	2·7	*
mean of rumen bacteria	*	9·3	2·6	*
fungi Penicillium notatum	38	4·0	1·0	1·3
algae Spirulina maxima	65	4·6	1·8	1·4

* unknown

way, the protein content per unit total mass is diminished. This can easily be avoided by culturing the organisms under conditions in which the growth is limited by the energy source. Subject to these variations, the percentage of crude protein is generally 50–55 for yeasts, 50–80 for bacteria, 15–45 for fungi and 20–60 for algae. Bacteria thus appear to be relatively rich in protein, fungi distinctly inferior and algae variable.

As protein is assumed to be 16 per cent nitrogen the crude protein content is usually taken to be 6·25 times the percentage of nitrogen, but in practice this is almost invariably an overestimate. One reason for this is that there is considerable nitrogen in materials other than protein (in purine and pyrimidine nucleic acid bases and in amino sugars in the cell wall). Again, there may be amino acids in cell-wall material which may not be available through normal digestion. Subject to the availability of more data it would be preferable to estimate protein content by the sum of the individual amino acids, although this procedure is tedious and liable to errors introduced by the presence of amino acids contained in unnatural peptides or cell-wall material.

The cell walls of the micro-organisms considered here are probably indigestible, being made of peptidoglycans, lipopolysaccharides, mucopolysaccharides or cellulose. The main significance of the wall is that it may protect the cytoplasm inside the cell from digestive enzymes and so reduce the availability of proteins contained therein. Experiments in rat-feeding studies show that rupturing *Bacillus megaterium* increases the digestibility of the SCP. Whether this is true for other sources of SCP remains to be established.

Although this has not yet been evaluated, the presence of unnatural stereo-isomers such as D-amino acids in the cell wall could reduce SCP nutritional value. D-Alanine and D-aspartic acid have recently been shown to have a growth-inhibiting effect in the chick, and D-alanine is a poorer source of non-essential nitrogen than L-alanine in the rat. If the cell wall is indeed indigestible, as has been suggested, the presence of these amino acids in SCP will present no problem since they are located exclusively in the wall.

Another factor which can affect the nutritional value of SCP is

the nucleic acid content. This can vary from 5–20 per cent of dry weight and tends to be highest for bacteria growing rapidly. It depends strongly upon growth rate so that considerable manipulation is possible. It is not yet clearly established whether intact nucleic acids can be digested by humans and, if so, whether they contribute to an unacceptably high level of blood uric acid. If desirable, reduction of the nucleic acid content should be relatively simple by permitting ribonuclease to hydrolyse it to soluble compounds which can be washed away.

The choice of a particular micro-organism depends upon a complicated balance of the various nutritional, engineering, food-technological and economic factors since the object is to produce not an ideal protein but an adequate and cheap one. The choice thus depends on local nutritional conditions, the availability of required raw materials and equipment, acceptability of the product by a given population, existence of the required technology, cost of the product and various governmental incentives.

In general, bacteria and yeasts seem more suitable than fungi and algae, which grow more slowly and are less efficient in converting either nitrogen or carbon into protein. Fungi and algae are also considerably lower in protein content and appear to have more marked deficiencies in essential amino acids than bacteria or yeasts. The cell wall of fungi and algae constitutes a greater fraction of the cell, and will presumably furnish only roughage. The problems arising from the production of toxic secondary products are at least as great as for bacteria and yeasts. The only significant advantage that filamentous fungi and algae possess is that they can be recovered from the fermentation broths by relatively simple operations such as string filtration, whereas for bacteria and yeasts more expensive means such as centrifugation will probably be required.

Higher fungi such as mushrooms constitute a desirable food, but as a protein source are likely to be far too expensive to consider on a worldwide basis. At present such fungi are produced by fermentation by at least one American firm at a price of approximately $6·60 per kilogramme dry weight and used mainly for flavouring.

The yield of protein per unit of substrate depends on the yield of cell mass per unit of substrate and the fraction of the cell mass that is protein. Generally, for yeasts or bacteria growing on sugars under conditions favourable to a high protein content a yield of cell mass of about 50 per cent can be obtained, but it is rare to achieve so high a level with fungi. For bacteria or yeasts growing on hydrocarbons yields can reach 100 per cent or even higher. This is accounted for by the low oxidation level of the hydrocarbon compared with the cell material.

On the other hand, while hydrocarbons thus yield a large amount of cell per unit of hydrocarbon utilized, the cells require much more oxygen than do cells grown on more oxygenated substrates. For a typical carbohydrate process 200 kg of sugar plus 70 kg of oxygen yield 100 kg of cell mass, whereas for a hydrocarbon process 100 kg of cell mass requires 100 kg of straight-chain paraffin but no less than 200 kg of oxygen. At current US prices the cost of various carbon sources per kilogramme of cell mass produced works out to about 44 cents for sucrose, 8·8 cents for molasses, 31 cents for soybean oil and 2·9 cents for kerosene or distillate oil. Thus the apparent advantage of the hydrocarbon over molasses is only about 6 cents per kilogramme of product, and would be even less in parts of the world where molasses is cheaper than in the US.

In some cases the oxygen requirement for hydrocarbon-grown cells may not be as high as this, but as a general estimate it is not misleading. Furthermore, as a result of this oxidation the heat produced by the hydrocarbon processes is double or triple that for the same weight of carbohydrate-grown cells. These two factors, greater oxygen requirement and increased heat production, tend largely to offset the savings realized from the use of the relatively cheaper hydrocarbon substrate. The cost of this additional oxygen is difficult to estimate accurately but, for a typical fermentor in which the extra oxygen is transferred from the air to the fermentation broth by increasing the agitation, the additional cost on this account of using a hydrocarbon substrate instead of a carbohydrate would be of the order of 1·3 cents per kilogramme of cells produced.

Moreover, the extra heat generated during growth on hydrocarbons poses another problem. The temperature of growth for

most bacteria, yeasts or moulds proposed for use as sources of SCP is in the range 25–40 °C. Maintenance of the best temperatures is relatively simple where plenty of cold (15–20 °C) water is available but, in many parts of the world, surface water temperatures during the hottest months are in excess of 30 °C, and even cooling towers may be ineffectual because of the high relative humidity. If mechanical refrigeration must be used to remove the heat produced, additional costs will be incurred of about 2·2–6·6 cents per kilogramme for hydrocarbon-grown cells and half as much for those grown on carbohydrate. Thus, the additional costs of supplying extra oxygen and refrigeration for a hydrocarbon process operated conventionally just about counterbalance the saving on the carbon source.

The costs of refrigeration could be saved if thermophilic organisms could be used, because these grow fastest at higher than room temperature. Until last year there had been no published reports of thermophilic bacteria or yeasts capable of growth on hydrocarbons, but we have recently isolated some bacteria capable of growth on alkanes at temperatures up to 70 °C. Not only would such organisms alleviate the need for fermentor cooling but they might also make it simple to operate without sterilization since relatively few organisms might be expected to grow on hydrocarbons at such a high temperature.

In the past year several industrial programmes have been described in which hydrocarbons are used as substrates for SCP production. British Petroleum, which conducted SCP research at Grangemouth, Scotland, and has a pilot plant at its French subsidiary's refinery at Lavera, has developed a process in which SCP animal feed is produced during removal of higher paraffins from crude oil, and another method based on growth on purified paraffin fractions. BP recently announced the start of construction of a £2 million plant at Lavera to produce 16 000 tonnes per year of SCP from gas oil; this should be completed by 1970. Esso, which has been engaged in a joint project with Nestlé, has concentrated on the production of SCP suitable for human consumption by using a purified paraffin substrate. Shell in Great Britain has announced work on a process using methane as the carbon source.

Most of these processes are probably directly applicable to practically every oil refinery. Although they obviously involve considerable capital expenditure on fermentors and separation centrifuges, it is always implied that the low cost of the carbon substrate is a key factor in the economics of the process. But in practice the substrate cost is not insignificant. Even methane has a US market price of about 2·2 cents per kilogramme. Where these substrates really are cheap is at the well site, but usually the other required nutrients and water are not available there or else the market is far away. Furthermore, domestic experience with the production of torula yeast from sulphite liquor and *Saccharomyces fragilis* from waste whey indicates that, even when the substrate is free or actually carries a credit for its use since alternative methods of disposal are more expensive, the resulting product still costs 22–33 cents per kilogramme of cells or about 66 cents per kilogramme of protein.

Altogether, although the cost of carbon substrate is not insignificant, neither is it a large fraction of the cost of the final product. In our opinion, the main advantage of using hydrocarbons as substrates for SCP is not their low cost but that they are available in very large quantities as non-agricultural products. Thus, their production will not vary with climatic variations; there will not be problems associated with the education of farmers; prices should be relatively stable; and the amount of hydrocarbon actually needed for SCP production should have little impact on the overall hydrocarbon-supply situation.

In addition to carbon and oxygen the manufacture of SCP requires a nitrogen source, minerals such as sulphate, phosphate and magnesium, and trace elements. Some strains may also require vitamins. With bacteria, yeasts and moulds the nitrogen source can conveniently be liquid ammonia while phosphate can be supplied as phosphoric acid and magnesium and other minerals can be supplied as inorganic salts. The quantitative requirements for these minerals and nitrogen sources will be approximately the same for all organisms. If vitamins must be furnished they should be supplied in the form of inexpensive complex materials such as molasses, distillers' solubles or corn steep liquor.

Growth rate is important but not nearly as crucial a factor as has been supposed. If the growth process is operated continuously –

and this is a reasonable assumption – the productivity is equal to the organism concentration multiplied by the specific growth rate. However, the process will be limited either by the maximum rate at which oxygen can be transferred from the air to the nutrient medium or by the transfer rate of the relatively insoluble hydrocarbon. The productivity will then be proportional to this transfer rate and if the growth rate of the cell is low then the organism concentration can be high, so that productivity can be constant regardless of the growth rates involved. This is true over a wide range of growth rates with the reservation that at low growth rates an appreciable amount of carbon source will be consumed for maintenance rather than growth, leading to low yield per unit of substrate. Low growth rates may also aggravate problems of contamination.

If the fermentor must be operated aseptically a cost is incurred in sterilizing the medium being introduced. Moreover, the system must be designed to keep out foreign organisms, and this again is possible but expensive. An alternative is to devise a set of culture conditions which favours the growth of one organism to the virtual exclusion of other possible contaminants. The desired result is not so much a product composed 100 per cent of one organism as to prevent the development in the fermentor of foreign organisms which can build up until they displace, or at least constitute a significant fraction of, the desired population. Possibly this can be done by operation at a high temperature or at a low pH value, or by choosing a substrate such as methane which is incapable of supporting the growth of all but a very few organisms. Much ecological work remains to be done to determine whether 'open' systems, as distinct from sterile and sealed ones, will lead to acceptable SCP.

Recovery of the product in a clean form is another necessary step. Part of the difficulty in the proposal to combine SCP production with petroleum refining is the necessity for washing every trace of residual crude oil from the cells. This is not necessary if a pure alkane or methane feed is used.

In any case the recovery of the cells, presumably by centrifugation, remains difficult and costly. Large or dense cells would be easier to collect than small cells covered with low-density residual hydrocarbon. Possibly they could be agglomerated by

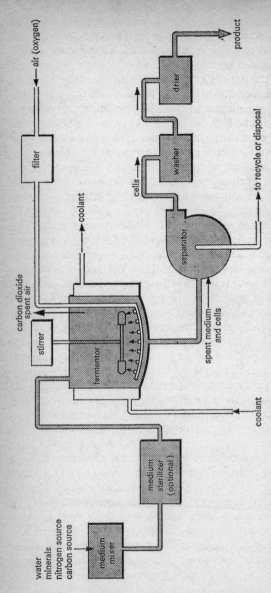

Figure 1 Production of a single-cell protein would follow a sequence broadly along the lines indicated in this simplified flowsheet. The input mineral would include such essential elements as potassium, phosphorus, magnesium and sulphur; the nitrogen source would be ammonia and the carbon source sugar or a hydrocarbon. The separator would be a centrifuge or other harvesting device and the spent medium emerging from it could be recycled if necessary to improve the yield. The final product would contain roughly 50 per cent protein measured as a proportion of its dry weight

adding flocculating agents. The cost of centrifugation will be inversely proportional to the concentration of cells in the medium, so it would be preferable to operate at a low growth rate and high cell concentration.

If algae are used to produce SCP many of the considerations relative to SCP produced on organic carbon substrates in fermentors still apply, but in altered form. In the growth of algae, carbon dioxide from the atmosphere or elsewhere is reduced and incorporated into the cell. The energy for this biosynthesis is obtained not from oxidation of an organic carbon source but from light. The apparent low cost of this process in which 'free light plus free carbon dioxide equals valuable protein' has stimulated considerable work on algal SCP.

Unfortunately, it appears that the low cost of carbon and energy is offset by other factors. In particular, since the process depends on photosynthesis, the layer of nutrient medium in which the cells grow must be relatively thin or most of it will not photosynthesize. Under these conditions the strong solar radiation in many countries will result in a high rate of evaporation which will require large amounts of cheap water. Furthermore, the concentration of cells obtained is not very high so that recovery costs are significant; and in the case of the green algae usually used, such as *Chlorella* or *Scenedesmus*, the amino acid pattern is not particularly good.

The Japanese are finding that these algae cannot be produced for less than about 84 cents per kilogramme and, although they are used in Japan for flavouring (not as a protein source), they are unpalatable to a Western taste. The possibility of using these algae for animal feed depends on producing them cheaply enough to compete with other cheap plant proteins. Although this may seem to be a dubious proposition it is conceivable that, if enough credit is given to water reclamation, the growth of algae on waste water might be economical and a group at the University of California at Berkeley are working on this.

A recent proposal of interest concerns the growth of the blue-green alga *Spirulina maxima*. The Institut Français du Pétrole has observed that, as we related at the outset, these rather large algae have been eaten for years by inhabitants of the Lake Chad and

Fort Lamy regions of the Sahara. They cling together because of their shape, so that they could be harvested by relatively inexpensive means such as a rotary string filter instead of centrifugation. Furthermore, their protein content and amino acid patterns are relatively favourable. Finally, whereas *Chlorella* has a distinctly unpleasant and sharp taste, *Spirulina* is rather inoffensive. These algae are certainly of sufficient interest to make further investigation worth-while, if only because they represent a different approach to the problem of SCP production.

The value of SCP as a protein supplement will be related not only to its nutritional qualities but also to the extent to which it can be used in conventional and non-conventional foods. This will depend on such factors as colour, flavour, solubility and compatibility with other ingredients of food products.

So far the food industry's experience in the use of SCP is limited. Its main uses in human food are as a flavouring material in yeast extract and dried torula as a source of B vitamins. The use of pressed yeast for food-processing operations which require a fermentation step does not fall into the SCP category.

In the manufacture of yeast extract, yeast is allowed to autolyse or is directly hydrolysed with hydrochloric acid. This process results in a soluble paste which can have a rather high salt content of up to 25 per cent. The autolytic process is potentially of interest but it is very slow and may not be applicable to large-scale production of an inexpensive product.

Dried yeast has been used extensively as a source of B vitamins, particularly in non-conventional food products. One example is the various formulations developed by the Institute of Nutrition for Central America and Panama [. . .] in which 3 per cent dried torula yeast is a standard ingredient. The use of microbial cells at such a low concentration poses little difficulty in formulation. If, however, the same material were to be used at much higher levels as a protein supplement the formulation problems are unknown. The various processing operations which might be applied to microbial cells to maximize both their nutritional and technological qualities are not well understood, but answers are being sought in a number of industrial and university laboratories.

How much the cells will have to be washed to provide a suitable

SCP will obviously depend upon such factors as the type of substrate used in the fermentation, the degree of odour or flavour which can be tolerated and the results of toxicological testing. Proposals to employ volatile organic solvents to remove trace quantities of residual hydrocarbon will probably not be very successful owing to the high cost of solvent handling and recovery.

Before cells can be used as a basis for SCP they must be killed. This again opens up a range of options and few data exist on which a choice of method for large-scale production can be based. Perhaps the first technique to be considered should be heating by means of some type of continuously operating heat exchanger. The critical parameter governing the choice of method is the proportion of viable cells that can be permitted in the final product, and this in turn depends on such considerations as the species of cell being produced and its ability to survive and reproduce in the human intestinal tract.

Since all SCP processing involves liquid phases, excess moisture must be removed, and once more a range of methods present themselves. Two ways of dehydrating cells are to dry them on a heated, rotating drum or spray them as a fine cloud in hot, dry air. In this case, choice of method depends upon the physical characteristics of the suspension and the desired properties of the final product. Clearly the drying conditions must not reduce the product's overall nutritional quality.

What additional type of processing might be applied to SCP to improve the value of the product? The improvement might be either in nutritional quality or in physical characteristics. We have previously postulated that one method of processing which might accomplish both these purposes is that of breaking the cell wall. This process has been shown to improve the nutritional value of *Bacillus megaterium* and of some yeast, but the improvement is so slight that its economic worth is doubtful.

This type of process might also be used for the improvement of the SCP's physical properties. A protein supplement, particularly for children, could be made in the form of a dry powder which can be rehydrated to form a beverage. Intact cells are difficult to maintain in stable suspension without the use of artificial stabilizers because the cells are insoluble and agglomerate into

clusters during heating and drying. But if they are fragmented before drying the proteins become soluble and the rate of settling of the insoluble broken cell wall becomes very slow. How cheaply this can be done is at present unknown. Producing protein concentrates by removing the cell wall entirely appears at present to be too expensive for immediate consideration.

Many problems must be solved before SCP can become a large-scale commercial reality, although none of these problems appears to require a great technological breakthrough. The overall cost of producing SCP cannot be greatly reduced by improvements in one specific process, but rather will require small gains in many areas.

The time scale for use of SCP in animal feed is quite short; in fact, commercial operations already exist for production of yeast from hydrocarbons in Taiwan and algae from sewage in the United States. But for use in human foods SCP would have to be substantially purer and freer from residual hydrocarbon contamination than is the case for animal feeds. As there are economic and governmental restrictions on the use of solvent extraction for removal of trace hydrocarbon residues, this higher standard may mean the fermentor will have to be charged with pre-purified alkane fractions. The proces of achieving an acceptably pure product might thus result in an unacceptable rise in cost. In any event the time scale for SCP for human consumption must be relatively long.

For SCP to have a significant impact on world nutrition would require the production and distribution of millions of tonnes of final product each year. If we are to take a lesson from history, specifically that of oilseed meals and fish-protein concentrates, the most important barriers to distribution will be those related to social eating patterns, government regulations and the technological status of local food industries in less-developed countries. It is reasonable to predict that at least the economic and technological problems of SCP production will be overcome within the coming decade. The torch will then have been passed to the social scientists, the marketing and distribution people and the statesmen.

28 A. B. Bowers

Farming Marine Fish

A. B. Bowers, 'Farming marine fish', *Science Journal*, vol. 2, 1966, no. 6, pp. 46–51.

Protein starvation is today one of the most serious human ailments; 80 per cent of the world's population receive less than thirty grammes of protein a day and about 10 per cent less than fifteen grammes. The minimum requirement is believed to be about thirty grammes and thus the world's total protein needs, for a population of 3000 million, is at least 33 million tonnes annually. However, in many technologically advanced countries – including the United Kingdom – between two and three times the minimum requirement is consumed daily; this means that, if protein starvation is to be eliminated, more than 100 million tonnes must be produced annually.

At present, 43 per cent of the available animal protein comes from meat, 35 per cent from milk and only 12 per cent from fish. In spite of this, more than half the world population depends solely on fish for their protein requirement. Furthermore, the fishing industry is much the most primitive of all protein-producing activities; it is still based on virtually indiscriminate capture, with little management of fish stocks. Freshwater fish, it is true, are cultured in ponds either on special fish farms or alternating with land crops such as paddy. The ability to use similar techniques on marine fish could, in theory, do much to improve the world's protein supply. Indeed, theoreticians have long recognized the need to exploit fish stocks in a more scientific manner. But in practice there has so far been very little research on marine-fish farming. The most important has been in Europe, and in the United Kingdom a vigorous if rather small programme has been under way for a number of years. Although concerned mainly with fish such as plaice and sole, I shall describe it in some detail in this article as the principles being elucidated in this

programme may later find worldwide applications if marine-fish farming proves to be an economic proposition.

Modern freshwater fish culture has developed from a centuries old tradition of rearing food-fish in ponds or artificial enclosures, and the techniques of rearing game-fish for stocking sport fisheries are well-established. Marine-fish farming, on the other hand, has hitherto made little progress because marine fish are difficult to rear and because stocks were generally believed to be inexhaustible.

Most marine food-fish produce thousands of eggs which float in the upper layers of the sea where many of them are eaten by marine animals. Hatching depends on the water temperature but usually takes about ten days. The newly hatched fish – larvae or fry – are rather weak and helpless but soon develop the ability to swim short distances and to feed on plankton. Most of the larvae die either because they fail to find food or because they are themselves eaten. Those that survive eventually metamorphose into adult fish. This is a complex process involving a change of shape, formation of a tough skin and scales and a change to more sophisticated physiological mechanisms. At the completion of metamorphosis flat-fishes, such as plaice and soles, take to the sea floor and many round fishes such as herring and mackerel gather in shoals; protective devices, such as camouflage and defensive habits, are developed and the mortality rate is much reduced. After metamorphosis many species of food-fish spend from one to two years in shallow coastal areas before migrating to the open sea.

In the late nineteenth century a few fishery biologists and administrators foresaw the danger of overfishing and looked for ways of increasing natural stocks. They reasoned that, since the most vulnerable stages of a fish's life were the egg and larva, nursing developing fish through the early stages would ensure that more would grow up to adult size. Experiments showed that millions of eggs could easily be collected from a few hundred fish kept in marine ponds and that these eggs could be incubated and hatched successfully using specially designed equipment. Unfortunately commercial hatcheries were built in the United States, Scandinavia and Britain before the experiments were completed. As a result

many millions of cod and plaice eggs were hatched but as techniques for mass-producing the right sort of food for the larvae were lacking, they had to be released into the sea at a very early age. Supporters of the hatchery movement, together with many fishermen, claimed that these releases increased the stocks but sceptics denied this and no convincing evidence was available. Gradually confidence waned, financial support was withdrawn and marine hatcheries closed down or were converted to fishery research stations.

Meanwhile Danish and British biologists had been experimenting with transplants of young wild plaice from the breeding grounds to areas where they were scarce. When the Danish experiments started in 1895 there were only a few migrant plaice in the broads of Lim Fjord and no resident spawning population. The young plaice were introduced when they were about seventeen centimetres long and grew quickly to edible size, sufficient numbers surviving to make large transplantations economically worth-while. For fifty years transplantations were made annually on a commercial scale except for a few interruptions during wartime, and over the first twenty-five years commercial catches in the broads were closely related to the number of transplants. Later there were some large variations in yield which were independent of the number of fish transplanted; this was thought to have been caused by natural immigration though, previously, this had been so small that it was unimportant. Over the whole period of fifty years the annual value of edible fish caught varied between three and seventeen times the cost of transplanting young fish.

In other Danish experiments North Sea plaice from coastal waters were transplanted into the Belt Sea and English workers moved live fish from coastal regions with a high population of young plaice to the Dogger Bank where there were few. In both cases the transplants grew remarkably quickly. Indeed, on the Dogger there was even some growth in length and weight during the winter, when coastal fish do not grow.

These experiments showed that live fish could be moved about without harm; that provision of adequate living space on good natural feeding grounds resulted in rapid growth; and that some local races of plaice had a very high growth potential.

In 1942 Dr F. Gross and his co-workers attempted to improve the amount of natural fish food produced in two sea-lochs in Scotland – one enclosed and one partially enclosed – by adding agricultural fertilizers to the sea-water. They applied 203 kg of calcium super-phosphate plus 610 kg of sodium nitrate or 5085 kg of ammonium sulphate per million cubic metres of sea-water [. . .] at intervals of about one month for two and a half years. The fertilizers stimulated the production of microscopic marine plants which supported an increased population of planktonic animals and so on up a 'food chain' culminating in the animals which form the food of fish.

As in both lochs the natural fish population was small, young plaice and flounders were transplanted to the smaller loch – many of them marked so that they could subsequently be recognized – while a hatchery was set up to produce plaice eggs and larvae to stock the larger loch. Wild flounders also migrated into it as the feeding improved. It was estimated that only 0·2 per cent of the plaice released as larvae survived, but in both localities the growth of the surviving plaice and flounders was very much better than in neighbouring unfertilized areas, thus demonstrating that marine-fish farming was technically feasible. Even so there were many critics because the experiments did not reveal the relationship between fish yield and the amount of fertilizer added, nor was sufficient information available for the economics of fish farming to be assessed.

These experiments also brought to light some unfortunate side effects and difficulties associated with fertilizing sea-lochs. Some of the fertilizer was taken up by large sea-weeds which grew luxuriantly but then died and rotted, producing poisonous gases – such as methane and hydrogen sulphide – and depleting oxygen in the water. There were difficulties caused by fresh water running into the lochs which reduced the salinity, and in some parts of the lochs the muddy bottom proved unsuitable for flat-fish. In the larger loch, which had an outlet to deeper water, it was found that as the fish grew they tended to emigrate and were lost.

In the past twenty years most of the problems connected with large-scale hatching and early rearing of flat-fish have been solved. A major advance had been made in 1938 when Gunnar Rollefsen

in Bergen found that marine-fish larvae would feed readily on newly hatched brine shrimps, but the application of this discovery had to wait till after the war.

The brine shrimp (*Artemia salina*) lives in salt lakes in many parts of the world. Each season large quantities of eggs about the size of fine sand grains dry out on the lake shores. These remain viable but inert, so they are easily packed, transported and stored. When the eggs are placed in warm salty water they hatch in two days, and the shrimp larvae can be used to feed larval fish. Experiments in Norway after the Second World War gave promising results with herring, cod, plaice and sole. Some were reared to metamorphosis, but the mortality between hatching and metamorphosis was high.

Between 1951 and 1962 J. E. Shelbourne and his co-workers at Lowestoft made steady advances in rearing techniques. Plaice eggs netted from the sea were hatched and reared in small glass tanks both at Lowestoft and at the University of Liverpool's marine station at Port Erin in the Isle of Man.

In 1962 Shelbourne treated eggs, spawned by some captive plaice, with antibiotics to control bacteria. Nearly all the eggs hatched into strong larvae and three-quarters of them were successfully reared into metamorphosed fish the size of postage stamps. This impressive achievement had two important results: metamorphosed fish, big enough and strong enough to survive transplantation, could now be produced in the large numbers required for critical experiments in fish farming; and for the first time fishery biologists had a successful standard technique against which to measure experimental changes in rearing conditions.

A great impetus was thus given to the fish-farming movement, and in 1963 the United Kingdom White Fish Authority and the Ministry of Agriculture, Fisheries and Food supported work on large-scale hatching and rearing of sea fish. A 'pilot-plant' hatchery was built at Port Erin which produced 160 000 metamorphosed plaice in 1964, 400 000 plaice and some soles in 1965. Its potential output is one million fish annually.

At Port Erin a parent stock of plaice is maintained in a large sea-water pond almost as large as a tennis court and eight feet deep. The fish are fed on boiled mussels or 'queens', or chopped raw herrings. Any addition to the stock of spawners is made

before the roes start to develop so that the fish have plenty of time to acclimatize to pond conditions before spawning. Spawning takes place from mid-February to the end of April, the eggs being gently skimmed from the surface of the pond with a large terylene net from which they are dipped out into a transfer vessel and taken into the hatchery.

The hatching and rearing tanks measure $1\cdot2 \times 0\cdot6 \times 0\cdot3$ m and are made of black polythene. They are arranged in racks in a room with the air cooled to 6 °C. For the first few days the eggs float in sea-water treated with streptomycin and penicillin to destroy any bacteria sticking to the egg membranes, and then a continuous gentle flow of cooled sea-water is run into the tanks. Three or four days after the eggs have hatched newly hatched brine shrimps are given daily as food. Very young larvae will eat ten newly hatched brine shrimps per day per fish, but by three months plaice will eat two hundred a day each, so at this stage it is more economical to use older brine shrimps. As the plaice grow, air and water temperatures are raised by $\frac{1}{2}$ °C per week, following the changes in the sea.

The number of plaice reared in each tank with $0\cdot8$ square metre of floor space has been as high as four thousand. Unfortunately plaice reared at this density show several peculiarities: their range in size is great but, on average, they are small for their age; many fail to develop pigments in some areas of their skin and look pie-bald; there are more fish with minor deformities than is desirable; and some become very aggressive towards one another. My own work has shown that these effects are associated with overcrowding and that, if the fish are given more room, they are larger, nearly all normally pigmented and have fewer abnormalities and injuries. While the fish are in the larval state they use the whole volume of water in the tank and many more can be kept without harmful crowding effects than when they start to metamorphose and spend most of their time on the tank floor. Available resources can therefore be used most economically by rearing the larval stages at high density and dispersing the fish to other tanks at about two months old.

Much of the current research on fish farming is directed towards determining how best to bring on the fish from postage-stamp size

to eatable size. Early ideas that this could effectively be done by releasing three- to four-month-old fish from the hatchery into the sea have largely been abandoned. The losses in natural populations have been found to be high from three months to one year old; 30 per cent or more die every month, which means that a stock of one million fish at three months old is reduced to 40 000 at one year old; losses in hatchery fish released into the sea were much higher than this in a pilot experiment undertaken recently.

More promising is the transfer of four-month-old hatchery fish to enclosed arms of the sea or to wholly artificial marine ponds. At Ard Toa in Argyllshire, Scotland, a small loch has been enclosed by sea-walls and inflowing streams have been diverted. This five-acre site is being used by the White Fish Authority, the Ministry of Agriculture, Fisheries and Food and Strathclyde University to study the engineering, chemical and biological problems associated with fish farming. These problems, many of which were indicated by Gross's loch-fertilization experiment, relate to control of water quality, particularly the maintenance of oxygen and salt content; control of predators such as crabs, which could enter the loch as larvae coming in with the sea-water and grow until they became a menace to young fish; finding economical methods of augmenting the natural food supply by the addition of fertilizers, and supplementing the diet of the fish with artificial foods if necessary.

Further field experiments are in progress at power stations which use sea-water for cooling. If the waste heat from the power plant can be used to maintain sea-water at spring or summer temperatures all the year round, high growth rates can be expected in fish kept in the water. Plaice usually stop growing in winter but if steady growth could be maintained throughout the year they would grow to marketable size in two to two and a half years as against four years under natural conditions. Here, however, there are difficulties with water quality; engineers like to chlorinate cooling water rather heavily to prevent the growth of marine organisms which cause blockage of the cooling system. The quantity of chlorine that fish can tolerate is small and either safe dilution factors must be used or a system of heat exchange between power-station water and clean sea-water will have to be worked out.

A. B. Bowers 331

In an attempt to find the ideal farm fish we have started to select strains of plaice adapted to domestication and characterized by rapid growth, good pigmentation and behaviour suited to tank conditions. This is necessarily a long-term programme because each generation of laboratory-reared plaice takes four years to reach breeding size, though it may be possible to shorten this time. It is also planned to make crosses between different geographical races of plaice, and to produce hybrids between plaice and flounder so that their suitability for farming can be tested. In nature, such hybrids are rare but they grow very quickly and are of excellent flavour and texture for eating. Rollefsen found, in 1940, that young hybrid plaice × flounder were more hardy than young plaice when they were kept together in a large sea-water pond. Flat-fish of greater economic value than plaice, such as soles, are already being reared and the number of species tested will be increased. Greater control of parent stocks and breeding conditions is being established. Thus, indoor ponds have been constructed in which plaice and soles have successfully spawned and, by controlling the conditions in these ponds, it may be possible to arrange the date of spawning to suit the biologists.

The key to the success or failure of fish farming lies in economic factors. I have reared plaice up to eatable size in the laboratory; the cost of food alone exceeded the market value of the fish, though it is now clear that these plaice were not given optimum conditions for rapid growth. At present it appears that in plaice an average of five pounds of food are needed to produce one pound of fish, though the conversion rates of some individual plaice are better than this. Thus, if the wholesale market price for plaice were $12\frac{1}{2}$ p per pound a fish food with a high proportion of protein must be found which costs not more than 2p per pound. For more expensive fish one could obviously afford to spend more on feeding them. Dr C. Nash of the White Fish Authority is working on the conversion rates obtainable with different foods, and different rates of feeding, to determine the most economical way of producing fish flesh in carnivorous species which eat animal foods. Meanwhile, Scottish scientists are looking for marketable species of marine fish which live wholly or partly on plant food which would be cheap to provide.

In the early stages of fish farming some saving in production costs could be obtained by 'fattening' wild fish caught at six months old in coastal waters and brought into enclosed lochs or concrete ponds. At a later stage the advantages of using specially bred strains of fish would probably offset the added expense of early rearing, but in the meantime much valuable experience would be gained. Several species of potential farm fish are known to be available in sufficient numbers to make a fattening programme worthwhile.

Research and development is now concentrating on the rearing of fish to eatable size; on hatching and rearing marine fish with a greater market value than plaice; and on the production of special strains of fish. In a few years time the costs of fish production as a commercial venture will be assessed. If it is demonstrated that farming of sea fish is, or could be, an economic way of producing food this would in no way be a threat to the fishing industry, which will continue to supply the bulk and variety of fish to the markets. Fish farming would be a useful supplement and would certainly have repercussions on fish processing and other industries. Fish processers could have a planned supply of a limited variety of fish when and at a size they required.

Agricultural feeding-stuffs manufacturers would no doubt turn their attention to the production of fish foods which could be supplied in a form that is easily handled, stored and weighed. Engineering techniques would be applied to the construction of fish-ponds over which control of water temperature, water quality and flow could be exercised. Efficient rearing of fish through the nursery stages calls for automation of the processes of food production, food distribution and tank cleaning; progress is already being made in applying electronic control to these routine jobs.

Biologists and technologists are working together on problems arising from fish farming and are finding the cross-fertilization of disciplines and techniques stimulating. This cooperation will be extended as fish-management techniques are evolved and demands are made on specialists parallel with those made by land farmers for advice on their problems. Farm animals today are very different from their wild ancestors; comparatively little is known

of the genetics of marine fish but there is no doubt that the potential exists for the breeding of productive, fast-growing strains. Control and treatment of disease and parasites will be necessary if large numbers of fish are to be kept in close proximity to each other.

Many scientists are responding to the challenge to their skill and imagination that is presented by the concept of farming sea fish. They now need the resources and time to develop successful ideas and techniques. I believe that it would be a great mistake to apply economic criteria too soon, or to be too optimisitc about the immediate future of sea-fish farming. It may be ten years before useful crops can be produced; it will be many more years before the full potential of the new science of marine-fish farming can be adequately assessed.

29 E. F. Knipling

Possibilities of Insect Control or Eradication Through the Use of Sexually Sterile Males

E. F. Knipling, 'Possibilities of insect control or eradication through
the use of sexually sterile males', *Journal of Economic Entomology*,
vol. 48, 1955, no. 4, pp. 459–62.

The purpose of this paper is to consider the possibility of controlling insects by releasing sexually sterile males among the existing natural population. The principles involved will be described and the potentialities as well as the limitations of the method as we know them at present, will be discussed.

The theoretical possibilities of insect control by utilizing induced sterility or other damaging effects to genetic material have been considered by the writer for a number of years. Until recently there was insufficient experimental evidence to justify applied research to explore the method for pest control or elimination. However, the recent paper by Baumhover and co-workers presented at the 1954 meeting of the Entomological Society of America reported marked progress in the cooperative experiment to eradicate the screw-worm, *Calitroga hominivorax* (Cqrl.), from the 170-square-mile island of Curacao, Netherlands Antilles. A. W. Lindquist of the Entomology Research Branch, US Department of Agriculture, announced the complete elimination of the insect on Curacao in a paper presented at the 1955 meeting of the Cotton States Branch, Entomological Society of America.

The experiment on Curacao was undertaken following three years of well-executed basic research under the leadership of R. C. Bushland of the Kerrville, Texas, laboratory of the Entomology Research Branch. Excellent cooperation was obtained from the Oak Ridge National Laboratory and the Netherlands Antilles Government.

Bushland and Hopkins published results of laboratory investigations which indicated the feasibility of studies such as those conducted on Curacao.

The studies by Bushland employing X-rays or gamma rays,

confirming earlier investigations by geneticists and cytologists demonstrated that screw-worm males (and females) could be made sexually sterile by exposing pupae to irradiation without serious adverse effects on the mating behavior of the insect. The research also showed that the female screw-worm fly normally mates once only during her lifetime, and if mated to a sexually sterile male her reproductive potential is completely destroyed. It was demonstrated, for example, that when sterile and fertile males in a ratio of four to one were introduced in a cage of virgin females about 80 per cent of the females deposited eggs that did not hatch. The investigation by Bushland also showed that the sterile males produced sperms that fertilized the eggs but the developing larvae did not hatch.

After these key questions were answered, it was necessary to devise special research procedures and to carry out difficult and complex investigations in the field to determine if the principles established in the laboratory could be confirmed under field conditions. Early field work in Florida (unpublished) was encouraging, but the Curacao experiments by Baumhover and co-workers fully established the soundness of these principles.

The success of the Curacao investigations has created great interest, and inquiries are being received as to possibilities of this method for the control or eradication of other insects. It is hoped that information in this paper will stimulate further research where indicated and at the same time discourage investigations where the sterility method obviously will not be feasible. The writer is of the opinion that this method will be difficult and costly under the most favorable circumstances, but it might prove practical primarily as an eradication tool for certain highly destructive pests or provide a method of preventing the build-up or spread of established infestations. Basic studies on the problem might also, in future years, lead to new and more economical ways to induce sterility and thereby extend the practicability of this approach to insect control.

How sterile males cause a population decline

Before discussing the various requirements in estimating the feasibility of the sterilization technique now known, it seems important to describe the way insect control is achieved through

the release of sterile males. The principle of reducing the population of an insect by the sustained release of sterile males among the existing population is a simple mathematical proposition. The effect can be expected to be rapid and dramatic if circumstances are favorable and if the results follow theoretical possibilities. To illustrate this, some hypothetical figures are presented. It might be stated that calculations of this kind formed the basis for further investigations on the screw-worm. The figures presented in Table 1 show the theoretical population trend of an insect in an area if sterile males are distributed in sufficient numbers to dominate the natural fertile male population initially by a ratio of two to one, and if the number of sterile males released is maintained at a constant level as the natural population decreases. It is assumed in this example that the sex ratio in the wild population is approximately equal and that the released sterile males compete fully with wild fertile males in mating with the existing virgin females. It is also assumed that the wild population is essentially stable and that maximum depression of the population in subsequent generations will result.

It is apparent from the figures given that the sustained release of fully competing sterile males in a stable insect population could have a marked and rapid depressing effect. It is realized that a completely stable population is seldom encountered. However, many economic insect species have developed fairly close equilibrium with their environment. The full biotic potential of an insect is seldom reached, because of predation or parasitism, diseases, use of chemical control measures, cultural practices, limited host materials, and other factors. During periods favorable for a marked increase in population of a species, a ratio of two sterile to one fertile male might be insufficient to cause a depression in population of the succeeding generation, even though such releases might substantially reduce the rate of increase. For this reason a higher ratio would probably be indicated in most cases. In my opinion nine sterile to one fertile insect should be released if possible to provide reasonable assurance of a downward trend in population of most insects. However, the optimum ratio will no doubt vary with the species, and the extent of normal population increase from one generation to the next. If initial release is properly timed to coincide with a natural or

E. F. Knipling 337

Table I

Assumed natural population of virgin females in the area	Number of sterile males released each generation	Ratio of sterile to fertile males competing for each virgin female	Percentage of females mated to sterile males	Theoretical population of fertile females each subsequent generation
1 000 000	2 000 000	2:1	66·7	333 333
333 333	2 000 000	6:1	85·7	47 619
47 619	2 000 000	42:1	97·7	1107
1107	2 000 000	1 807:1	99·95	Less than 1

induced downward population trend, a low ratio of sterile to fertile insects might depress a population near the theoretical maximum. Subsequent releases at the same rate for one or two generations might then result in a sufficiently high ratio of sterile to fertile insects to continue the downward population trend even though conditions by that time, in the absence of sterile insects, would be favorable for a substantial increase. In order to determine the potential value of released sterile insects in control or eradication programs, population densities and population trends under various circumstances must be considered.

In the screw-worm eradication experiment on the island of Curacao the results were surprisingly close to maximum theoretical population depression. Within about three months after the eradication attempt was started, or in about three generations, the screw-worm was eliminated. The exact ratio of released sterile males to wild fertile males at the start of the experiment is not known, but it probably ranged between two and four to one.

Factors to consider in appraising feasibility of insect control by releasing sterile males

On the basis of the principle presented above we might list the factors to consider in appraising the possibilities of the method for controlling or eradicating an insect under various conditions.

1. An economical method of rearing millions of insects must be known or capable of development.

2. The insect must be of a type that can be readily dispersed so that released sterile males will be about as accessible to the virgin females in nature as are the competing fertile males.

3. The irradiation or other sterilization methods must produce sterility without serious adverse effects on the mating behavior or length of life of the males.

4. Females must normally mate only once. If females of a species mate more frequently, the sperms from irradiated (sterile) males must be produced in essentially the same number and compete with sperms from fertile males.

5. The insect to be controlled must have a low inherent population, or the species must under natural or induced circumstances reach a population sufficiently low to make it economically feasible to

rear and release enough sterile males to effect a further downward trend in the population of subsequent generations.

In estimating the possibilities of releasing sterile insects of any given species, one must find the first four conditions favorable before considering the fifth. Failure to meet any one of the first four requirements will establish the impracticability of this procedure for insect control. However, if the first four requirements can be met, the practicability of the method will depend on the population density and the number of sterile insects that will be required to effect a downward trend in the population. The possibility of releasing a dominating sterile population might be considered for three types of circumstances.

(a) *For controlling important pests that are normally present in small numbers.* The screw-worm and the Australian sheep blow-fly are examples of species where it might be economically feasible to rear and release sufficient sterile males to effect a downward trend in the population. Cattle grubs, *Hypoderma* spp., and other bot flies are present in relatively low numbers, but rearing methods may never be developed. It is possible that hornworm moths and tsetse flies in certain areas are present in sufficiently low numbers provided mass-rearing methods can be developed.

(b) *For newly established insect infestations before the infested area becomes large or the population density becomes high.* The Mexican fruit-fly infestation now present along the California–Mexico border is a type of incipient infestation where the method might conceivably be economical and effective.

(c) *As an adjunct to other control methods.* Several possibilities might be considered under such circumstances. In a small but intensive crop-production area special control efforts through the use of insecticides, biological control agents, or cultural practices might be employed to reduce an insect population to a low point, thus making it practical to rear and release a preponderance of sterile males.

The possibility of such procedure might be considered for various insects in which the cost of eradication would justified.

For example, in connection with the attempted eradication of *Anopheles* mosquitoes on Sardinia, a program sponsored by the Rockefeller Foundation, the use of insecticides reduced the population density of certain species to a point where it was difficult to find specimens, but complete elimination was not achieved. The regular distribution of a large number of sterile males in the area for several generations after the population had been greatly depressed might have destroyed the reproductive potential of the few females that escaped destruction by the insecticides.

Undoubtedly the intensive use of insecticides against such pests as the codling moth in areas such as the Yakima Valley in Washington reduces the population to a low density. If economical mass-rearing methods at a cost comparable to that of the screwworm or certain fruit flies were available, and if other circumstances should prove favorable, it might be practical to release sufficient sterile males in the spring to prevent the normal seasonal build-up. Releases for two or more seasons might eliminate infestations or reduce them to a point where routine releases to prevent economic population levels could be made more economically than spray schedules. Chances of success of such a procedure might be improved if a strain highly resistant to an insecticide could be reared, sterilized, and released in conjunction with regular spray schedules in the early stages of such program.

Discussion

It has not been demonstrated that the eradication of the screwworm will be economically feasible in any of the infested areas in the United States. We know, however, that the principle is sound so far as achieving eradication at a cost that approaches practicability. Further research on rearing techniques and release methods will no doubt reduce the unit-area cost below that in the Curacao experiment. It therefore seems worthwhile for investigators to give some thought to this method of controlling other pests.

The possibilities discussed in this paper are admittedly highly speculative. As pointed out by Bushland, considerable information has been obtained by various investigators on the effects of ionizing irradiations on insects. Other workers, including Hassett and Jenkins, Plough, and Sullivan and Grosch, have added to our

knowledge of the subject. However, we know very little about the mating habits of most insects or the effects of gamma radiations on the activity of sperms. We have very little information on the variations in population density in terms of actual numbers in a given area or on the factors governing their biotic potential. Factors other than those mentioned must be considered. It probably would be impractical to release insects which are highly destructive in the adult stage. For insects having long life cycles the procedure would probably be too slow for practical consideration. It would not be realistic to enter into a costly eradication program unless practical ways of preventing re-establishment could be put into effect or unless it would be economically feasible to continue indefinitely releases in sufficient numbers of prevent a buildup of damaging numbers.

The problem of segregating sexes of an insect to avoid damage by the females or to increase the effectiveness of the released males has not been mentioned. This matter has been given considerable thought in connection with the screw-worm research project. Bushland and Hopkins showed that females of the screw-worm can also be made completely sterile. They also have shown that the presence of sterile (irradiated) females in caged populations of fertile and sterile males and unmated fertile females does not significantly reduce the number of unirradiated females producing sterile egg masses. It may be highly significant also that the gamma-irradiated females are not capable of depositing eggs, although they may attempt to oviposit.

Mass rearing and release of parasites as a means of controlling destructive pests is currently being given more consideration by entomologists. The release of parasites under conditions where there is an abundance of host material may result in a rapid increase in parasites of the following generation, thus increasing the ratio of parasites to hosts. This is no doubt an advantage over the male-sterilization procedure for reducing the population of certain insects. When the host insects are greatly reduced, parasites become less efficient. In some respects a sterile male of the species to be controlled can be regarded in the same light as a parasite. If a marked downward trend in the population is achieved by releasing sterile insects, and if such releases are maintained at a constant level, the ratio of sterile to fertile insects in-

creases rapidly. Perhaps the greatest advantage of the sterilization technique over mass liberation of parasites is the probability that sterile males, because of the mating instinct, would seek out and destroy a few surviving females more effectively than the most efficient parasites known. The mass release of parasites or predators followed by the release of sterile males might be the most practical way to achieve control for certain pests.

Those who wish to consider employing sexually sterile insects as a means of insect control by the procedures discussed should weigh carefully the many limiting factors before expending much research effort on such a problem. If the production and release of sterile insects might conceivably aid in control or eradication programs, exploratory research as a guide for further investigations would seem justified. Irradiation equipment, as described by Darden, which was employed in the screw-worm research, may prove helpful in such preliminary investigations.

From a long-range viewpoint there is much that should be done to explore the feasibility of inducing sterility in insects in some manner among field populations. If practical ways could be developed by treating the natural population, thus obviating the necessity for rearing and distributing sterile insects, the feasibility of the method might be extended for a wide range of pests. For certain insects a chemical treatment to induce sterility instead of death would have great advantages over conventional insecticides. If, for example, a chemical spray or dust could be developed which would induce sterility instead of death in both sexes of the boll weevil when routinely applied to cotton, the treatment would not only destroy the biotic potential of the generation exposed but the infertile males would destroy the fertility of females which inevitably escape treatment. The addition to house-fly baits of a chemical which would induce sterility would in time be more effective in fly control than a lethal toxicant. If a mobile irradiation unit could be developed which would cause sterility instead of death for all exposed pink bollworms in the cotton fields after harvest, such equipment would theoretically be far more effective in control or eradication programs than a similar device which would actually kill the insect.

Summary

Research on the screw-worm, *Callitroga hominivorax* (Cqrl.), has demonstrated that the release of large numbers of sexually sterile males in an insect population will reduce the existing natural population. The soundness of this principle as a potential means of controlling certain insects has been established. A number of key factors must be considered and resolved, however, before the procedure can be regarded as feasible for eradicating or controlling any given pest.

1. A method of mass rearing of the insect must be available.

2. Adequate dispersion of the released sterile males must be obtained.

3. The sterilization procedure must not adversely affect the mating behavior of the males.

4. The female of the insect to be controlled must normally mate only once, or if more frequent matings occur the sperms from gamma-irradiated males must compete with those from fertile males.

5. The population density of the insect must be inherently low or the population must be reduced by other means to a level which will make it economically feasible to release a dominant population of sterile males over an extended period of time.

Research to develop ways to induce sterility instead of death among field populations of pest species and the advantage of this approach over lethal measures is stressed.

Further Reading

Cellular production lines

B. Afzelius, *Anatomy of the Cell*, University of Chicago Press, 1964.
A very readable account of cell biology.

J. A. V. Butler, *Gene Control in the Living Cell*, Allen & Unwin,
London, and Basic Books, New York, 1968.

J. A. V. Butler, *The Life Process*, Allen & Unwin, London, and Basic
Books, New York, 1970.

D. Chapman and R. R. Leslie, *Molecular Biophysics*, Oliver &
Boyd, 1967.
A little more difficult than the above.

D. W. Fawcett, *An Atlas of Fine Structure: The Cell*.
A magnificent collection of electron micrographs with explanatory
text.

J. L. Howland, *Introduction to Cell Physiology*, Macmillan, 1968.
A comprehensive review with more attention to cell biochemistry
than the previous book.

A. L. Lehninger, *Bioenergetics*, Benjamin, 1965.
Emphasizes the importance of energy flow in biological systems.

J. Paul, *Cell Biology*, Heinemann, 1965.
A useful elementary introduction to the topic.

J. Ramsay, *The Experimental Basis of Modern Biology*, Cambridge
University Press, 1965.
One of the most important books at A-level and first-year
university level for many years. Should establish biology as a
multidisciplinary science very firmly in your mind.

C. H. Waddington, *The Nature of Life*, Allen & Unwin, 1961.

H. R. Wilson, *The Diffraction of X-rays by Proteins, Nucleic Acids
and Viruses*, Arnold, 1966.
A relatively simple account of modern physics and chemistry, and
how they affect certain aspects of biological investigation.

Processes and their control in the organism

S. A. Barnett, *A Study in Behaviour*, Methuen, 1963.
Comprehensive study of individual and social behaviour of rats.

W. M. M. Baron, *Organization in Plants*, Arnold, 1963.

W. M. M. Baron, *Water in Plant Life*, Heinemann, 1967.
Two books which deal in a readable fashion with the physiology of plants.

J. D. Carthy, *Animal Behaviour*, Aldus Books, 1965.
A magnificent introduction to this branch of modern biology. Illustrated with many fine photographs.

A. G. Clegg and P. C. Clegg, *Biology of the Mammal*, Heinemann, 1962.
A well-established text which deals with mammalian organization at several functional levels.

G. E. Fogg, *The Growth of Plants*, Penguin, 1963.
Fascinating description of plant growth in all its aspects.

J. E. Harker, *Physiology of Diurnal Rhythms*, Cambridge University Press, 1964.
A good introduction to the topic, providing a wide range of material with good diagrams.

G. M. Hughes, *Comparative Physiology of Vertebrate Respiration*, Heinemann, 1963.
A description of the mechanics and chemistry of respiratory processes from fishes to mammals. Good clear diagrams are a feature of this book.

R. O. Knight, *The Plant in Relation to Water*, Heinemann, 1965.

A. M. Lockwood, *Animal Body Fluids and their Regulation*, Heinemann, 1963.
Two useful books which are complementary.

G. Marshall and G. M. Hughes, *Physiology of Mammals and other Vertebrates*, Heinemann, 1965.
A very readable, comprehensive introduction at A-level.

Structures, processes and control in populations
B. G. Ashton, *Genes, Chromosomes and Evolution*, Longman, 1967.
As its title suggests, presents a genetic view of the evolutionary process.

J. T. Bonner, *Cells and Societies*, Princeton University Press, 1955.
An interesting book, drawing together some of the ideas presented in parts 1 and 3 of this book of Readings.

T. O. Browning, *Animal Populations*, Allen & Unwin, 1963.
This book presents an idea of the dynamics of populations which

is different to those considered by Perrins in the extracts from his paper presented in this book. Rather on the difficult side.

A. J. Cain, *Animal Species and their Evolution*, Hutchinson, 1963.
Evolution looked at by a taxonomist.

W. H. Dowdeswell, *Practical Animal Ecology*, Methuen, 1963.
A must for anyone interested in this topic. Explains the techniques used in some of the Readings.

H. Frings and M. Frings, *Animal Communication*, Blaisdell, 1964.
Background information to the papers presented here which deal with this topic.

R. McArthur and J. Connell, *Biology of Populations*, Wiley, 1966.
A look at the broader aspects of population study but also reaches some depth, too. Perhaps not for beginners.

M. Hogg, *A Biology of Man*, Heinemann, 1963.
An introductory text which deals with some of the issues raised in this part of this book.

D. Lack, *Population Studies of Birds*, Oxford University Press, 1966.
For the ornithologist, this is certainly the book to use as a comprehensive review of population dynamics, but again, not until you have read a more elementary text.

J. Phillipson, *Ecological Energetics*, Arnold, 1966.
A fascinating little book which is a very clear exposition of what can be a complex series of concepts.

P. M. Sheppard, *Natural Selection and Heredity*, Hutchinson, 1959.
Along with the other books on the topic on this list, a good exposition of population genetics and evolution.

G. M. Smith, *The Theory of Evolution*, Penguin, 1958.
Evolution in its widest sense is expertly dealt with in this inexpensive edition.

Finally, two books which are not specifically related to the subject matter of this part, but which are related to the biology of man in its widest implications and, I think, essential reading for all educated persons today whether biologist or non-biologist:

J. K. Brierley, *Biology and the Social Crisis*, Heinemann, 1967.

G. R. Taylor, *The Biological Time Bomb*, Thames & Hudson, 1968.

Glossary

Adsorption Condensation in the form of a film of molecules of a gas or of a dissolved or suspended substance upon the surface of a solid

Allopolyploid A polyploid in which the sets of chromosomes come from different species

Ambient Surrounding

Anisotropic Possessing different physical properties, e.g. refractive index, in different directions

Antigen A substance capable of stimulating the formation of an antibody

Anuran Amphibians of the subclass Anura, i.e. the frogs and toads

Artefact The result of human interference in the natural world, e.g. the production of apparent structures within cells by the process of staining

Assay Estimation, especially of chemical content, reactivity, etc.

Autorhythmic Capable of generating regular nervous impulses without outside stimulation

Autopolyploid A polyploid formed of two or more sets of chromosomes from the same species

Autosome A chromosome other than the sex chromosomes, X and Y chromosomes

Biomass The amount of material in a species or population of living organisms

Carotenoid Orange pigments found in many plant tissues

Catalysis Alteration of the rate of a chemical reaction by chemical means, in living organisms by the use of enzymes

Chelate A chemical compound in which a single ligand (e.g. a metallic ion) is bound to the rest of the molecule at more than one place

Chromatin A protein constituent of chromosomes

Chromatin-positive Carrying more than one X chromosome, which occurs as a 'sex chromatin body'

Chromosome mosaic A body composed of cells containing at least two different types of chromosome complement

Colloid A substance divided into particles larger than the dispersed phase of a true solution, yet smaller than the particles of a

suspension. Frequently the particles acquire an electric charge which confers special properties on the colloid system. Many clays are colloids

Conditioning A primitive form of learning in which a reflex response becomes associated with a new stimulus. Hence *Conditioned response* – a response which has become associated with a new stimulus

Demographer A person concerned with the scientific study of population changes

Eclosion The emergence of an adult insect from a pupa
Endogenous Originating within a tissue, or organ
Exogenous Originating outside a tissue or organ, i.e. requiring an external stimulus
Exteroceptor A sense organ concerned with the reception of stimuli from outside the organism
Extrapolate Using a set of data to make inferences about conditions outside the range of the data
Extinction (when used in connection with learning behaviour) The loss of a response which had previously been learnt

Feedback The use of the products of an action to control the input
Flux The movement or flow of a fluid in a field. Can also refer to magnetic force

Geiger counter (perhaps more correctly a Geiger–Müller counter) An instrument for detecting ionizing radiation

Haploid The number of chromosomes carried by a mature gamete. This is usually half the number in a somatic cell nucleus which has the diploid number of chromosomes
Helical A spiral coil shape
Homogenate A substance of like consistency throughout, containing no separate structures within itself

In vitro (literally 'in glass') In laboratory vessels
In vivo In the living organism

Latent period The time between the application of a stimulus and the response
Luciferase An enzyme involved in the processes by which organisms can emit light

Matrix (in the sense used in van Bergeijk's paper) A table composed of rows and columns in which the data is assembled for analysis

Mitochondrion Cellular structures in which the energy-releasing reactions appear to occur

Moiety A part, properly a half, of some structure

Mongol A particular type of mental deficiency resulting from mistakes in cell division

Monogastric Having only one stomach (as opposed to ruminants which have several 'stomachs')

Multilamellate Consisting of many layers

Myoglobin A protein which acts as an energy store in muscle

Neuro-endocrine Referring to the internal signalling system within an animal. The nervous system and the various ductless glands are very closely interlinked in several ways, and should perhaps be thought of as a single system

Nidification Nest building

Parameter The measurable qualities of a population or object which describe it accurately. Parameters sometimes cannot be known accurately, but only estimated from samples

Parthenocarpically Formed without fertilization

Pathogen A parasite which causes disease

Photochemical A chemical reaction, the rate of which is affected by the action of light

Photometer A device for measuring the intensity of light

Photomorphogenic Describes a process in which the form of the finished product is controlled by light

Photoperiod The length of exposure to light

Photostimulation Stimulation by light

Polarity Having two distinct ends; in a molecule, it refers to the electric charges on different ends of the structure

Polymer A molecule built up of repeated units

Polyploid Having three or more times the haploid number of chromosomes. Hence *tetraploid* – four sets of haploid chromosomes, and *hexaploid* – having six sets

Potential Rather a difficult concept to explain exactly, but a difference of electric potential is the cause of the flow of electricity from one point to another

Progenitor Ancestor

Purkinje fibre A system of modified cardiac muscle cells in the intraventricular septum which conducts the impulse through the heart

RNA A nucleic acid containing the sugar ribose. Occurs in several different forms (messenger RNA, transfer RNA) in a cell, each with a different function

Refractory Non-excitable

Releaser A behavioural action, or a chemical substance which 'releases' appropriate behaviour in another organism

Resonance hybrid The 'average' structure of a molecule whose properties and bonding cannot be described by any one distribution of electrons

Ribosome A structure within a cell on which protein synthesis takes place

Running average A statistical technique for revealing trends in which averages for partially overlapping periods are taken, e.g. January–February–March; February–March–April; March–April–May; etc.

Sib or *Sibling* Brother or sister

Signal-to-noise ratio (a term borrowed from electronics) The extent to which 'real' effects are seen or heard against random variations in the background

Sinusoidal A graph of an electric potential against time may start from zero, rise to a maximum and then decrease, passing through zero, to a minimum before rising again. It is the shape of a typical sine wave

Stereoisomers Molecules of the same compound containing the same number of atoms, but in which the atoms are arranged in different ways (often mirror images of each other)

Steroids A group of hydrocarbon compounds containing a series of ring structures. Have a wide range of biological uses

Sublime The conversion of a vapour directly to a solid without passing through a liquid state

Substrate The layer underneath

Thermophilic Heat-loving

Triploid Having three sets of haploid chromosomes

Unconditional response A reflex response to a natural stimulus, e.g. salivating at the sight of food

Vulcanism Volcanic action

Acknowledgements

In preparing this selection of papers I have been helped considerably by the library staff at the City of Leeds and Carnegie College of Education, especially Miss Jennifer Platts, who has rarely failed to produce the copy of a journal required within minutes of being asked. Miss Shirley Ashton produced an impeccable typescript from my assemblage of tatty typing, scribbled notes and sticky tape, and to her I am eternally grateful.

Mr Peter Kelly, of the Nuffield Foundation A-level Biology project, kindly furnished me with a copy of the recommended background reading for their courses. This list gave some useful leads, and some of their recommended papers have been included.

Permission to reproduce the Readings in this volume is acknowledged to the following sources:

1 *Science Journal* and D. Chapman
2 National Academy of Sciences and A. L. Lehninger
3 *Science Journal* and P. Echlin
4 *New Scientist*
5 *New Scientist*
6 *Science Journal* and J. A. V. Butler
7 *New Scientist*
8 University of Chicago Press
9 *Science Journal* and A. C. T. North
10 English Universities Press and W. H. Freeman
11 *New Scientist*
12 Zoological Society of London
13 The British Association for the Advancement of Science, and M. B. Wilkins
14 The British Association for the Advancement of Science, and D. Noble
15 *Animal Behaviour* and W. A. Van Bergeijk
16 The British Association for the Advancement of Science, and W. M. Court Brown
17 *Nature*, E. H. Hazelhoff and H. H. Evenhuis
18 *New Scientist*
19 *Animal Ecology* and C. M. Perrins
20 *Science Journal* and R. Revelle

21 *Science Journal* and B. P. Moore
22 *New Scientist*
23 *Animals* and M. P. Harris
24 *Science in Action*
25 *Science in Action*
26 The British Association for the Advancement of Science, and R. Riley
27 *Science Journal*, S. R. Tannenbaum and R. I. Mateles
28 *Science Journal* and A. B. Bowers
29 Entomological Society of America

Index

Modern Chemistry
Edited by J. G. Stark

These Readings reflect the many profound changes chemistry has seen
in the twentieth century, and provide clear and stimulating
introductions to the most important new developments. The editor
has made his selection and provided editorial introductions for the
more advanced student at school and the student beginning at
University. The major areas in which papers have been selected are
atomic and molecular structure, stereochemistry, energetics and
kinetics, acids and bases, the chemical elements, and organic reaction
mechanisms. A number of suggestions for further readings are provided
at the end of each article.

Many of the papers were not originally published in SI units. To bring
the book into line with current thinking on the subject, the editor has
provided a note on SI units and given the relevant conversion factors.

Modern Physics
edited by David Webber

The world of modern physics has become so specialized and is advancing so rapidly that it is difficult to appreciate its scope. For a student intending to take his studies to a more advanced level, David Webber provides, in this edited collection of Readings, a sample of some of the more important and basic areas of modern physics. The papers have been chosen because their style is approachable and because they do not involve very sophisticated mathematics. The editor, in his part introductions, provides a useful background summary of work which the papers assume. This allows the reader to enter easily into the five main areas discussed in this book: Quantum Theory, Nuclear Physics and Fundamental Particles, The Solid State, Plasma Physics, and Relativity. The author has provided a list of further reading and has converted the units to conform to the SI system.

The Chemistry of Life

Steven Rose

The molecular structure of a protein (insulin) was described in detail for the first time in 1956: today such procedures are routine. Not only has the pace of biochemistry accelerated in recent years: with the perfection of the electron microscope and the development of cybernetics, the science has also widened and grown more complex.

The Chemistry of Life outlines the scope and achievement of a science which began as the study of the chemical constituents of living matter. Dealing successively with the chemical analysis of the living animal cell, the conversions induced between chemicals by the enzymes acting as catalysts, and the self-regulating nature of cells, Professor Rose explains how the design of particular cells influences their functions within the living organism as a whole.

Biochemistry is a difficult subject. But it is presented here as simply as accuracy will permit by a young research chemist who conveys much of the adventure of discovery implicit in a science which may one day answer the eternal question: 'What is life?'

The Structure of Life

Royston Clowes

Since the development, during the 1940s, of the electron microscope, with its vastly increased definition, a revolutionary new science has been born – that of molecular biology. As C. P. Snow has said: 'This branch of science is likely to affect the way in which *men think of themselves* more profoundly than any scientific advance since Darwin's – and probably more so than Darwin's.'

Molecular biologists study the structure of cells, the composition of proteins, and even the nature of viruses in terms of molecular chemistry. They explain hereditary mechanisms, for instance, by the behaviour of molecules of nucleic acids and of the proteins whose synthesis they control. In time they may place within man's grasp the key to life's fundamental processes and even to his own evolution.

In this introduction to a science which could prove more potent than atomic physics, Professor Royston Clowes outlines the latest interpretation of the living processes and foresees the passing of such axiomatic concepts as youth and age, health and disease, male and female, life and death.

Introducing Biology

James F. Riley

Biology, the science of life, embraces the growth, structure, and functioning of organisms, animals, and plants, the how and why of our bodies, and man's position in the universe of nature. And recent advances in biology have been no less dramatic than in other branches of science. Aided by the electron microscope the experimental biologist, long content just to analyse the living processes, is now poised for the artificial synthesis of life.

In this brief introductory outline of our knowledge of man and his environment, Dr Riley's themes are the workings of evolution and natural selection. He employs the lives and characters of the great biologists to disclose the progress of discovery in these fields, and by means of this historical approach builds up a stimulating (and painless) introduction to biology, which will appeal to career and leisure students alike.